FRESH WATER

RELATED BOOKS BY E. C. PIELOU

A Naturalist's Guide to the Arctic

After the Ice Age: The Return of Life to Glaciated North America

(both published by the University of Chicago Press)

The World of Northern Evergreens

FRESH WATER

E. C. PIELOU

The University of Chicago Press
Chicago & London

The University of Chicago Press, Chicago 60637
The University of Chicago Press, Ltd., London

Printed in the United States of America
15 14 13 12 11 10 09 08 07 06   2 3 4 5 6
ISBN: 0-226-66815-0 (cloth)
ISBN: 0-226-66816-9 (paper)

Library of Congress Cataloging-in-Publication Data

Pielou, E. C.
Fresh water / E. C. Pielou.
p. cm.
Includes bibliographical references and index.
ISBN 0-226-66815-0 (alk. paper).
1. Hydrology. I. Title.
GB661.2.P54 1998
551.48—dc21   97-51562
CIP

for PATRICK

Contents

Prologue

When color photographs of the earth as it appears from space were first published, it was a revelation: they showed our planet to be astonishingly beautiful. We were taken by surprise. What makes the earth so beautiful is its abundant water. The great expanses of vivid blue ocean with swirling, sunlit clouds above them should not have caused surprise, but the reality exceeded everybody's expectations. The pictures must have brought home to all who saw them the importance of water to our planet.

The satellite photos aroused a new appreciation of water in its many aspects. Three kinds of water dominate the picture: oceans, clouds, and the ice caps of Antarctica and Greenland. Not obvious in the photos, but there all the same, is the fresh water of rivers, streams, lakes, ponds, and wetlands, not to mention underground aquifers.

The pictures remind us of the fresh water even though they show only a little of it. They emphasize that, although water as a whole is abundant, usable water of the kind needed by humans is not. Considered thoughtfully, the pictures make it believable that a shortage of clean fresh water could well be the ultimate limiting factor to human population growth. Realization of the growing water shortage has led to justifiable concern about the inadequacy of the world's water resources for the growing population; nowadays many people think of fresh water *only* in the context of its availability as a "natural resource" for humanity.

This is a pity, because fresh water is much more than a "natural resource"; it is an integral part of the planet. It has been here for billions of years, and will continue to be here for billions to come, long after our species has disappeared. It is part of nature, a dynamic, functioning part. The liveliness of a sparkling mountain stream, the inexorable flow of a big river, the roar of a waterfall, and the silent tranquility of a calm lake are familiar examples of the sights and sounds of fresh water that add immeasurably to our appreciation of the natural world. Everyone is conscious of the way fresh water adds to the enjoyment of life, but not many people study it. The subject has been left to professional hydrologists in spite of its obvious interest to observant naturalists; it has been unaccountably neglected by amateur scientists.

Fresh water as nature made it is all around us, in rivers, lakes, and wetlands, some of them still pristine; as hidden groundwater that bubbles to the surface in springs; as invisible water vapor in the air becoming apparent when it forms clouds; as rain, snow, and ice. Living things depend on water but water does not depend on living things. It has a life of its own.

1

The Water Cycle

The most noteworthy characteristic of any small body of fresh water—be it a pond, a stream, an icicle, or a rain cloud—is its impermanence. Ponds evaporate, streams flow to the sea, icicles melt and dribble away, rain falls: water is forever on the move, repeatedly changing its state—liquid, ice, or vapor—in the process. In a word, it is dynamic. In much the same way that every living organism has a life cycle, water has a water cycle: it circulates. Indeed, all the water on earth is constantly circulating.

Figure 1.1 represents the cycle. In one form or another it is familiar to almost everybody. Water falls to earth as rain or snow, some on land and some at sea; much of what reaches the land drains into the sea, by surface routes or below the ground. The rest evaporates back into the atmosphere, either directly or through the vegetation: all plants on the land absorb water from the ground and exhale most of it into the air, as vapor, in a process called *transpiration*. The vapor from the land mixes with vapor evaporating from the sea, and, together, they condense into rain-giving clouds again. The cycle keeps on indefinitely, with the world ocean as the reservoir. The bulk of all this water is therefore salty sea water. Only the water temporarily withdrawn from the ocean is fresh.

Now for some statistics. The total amount of water on earth at the present time is approximately 1.4 billion cubic kilometers,[1] an impossibly large quantity to visualize. If it were solidified into a cube, each edge of the cube would

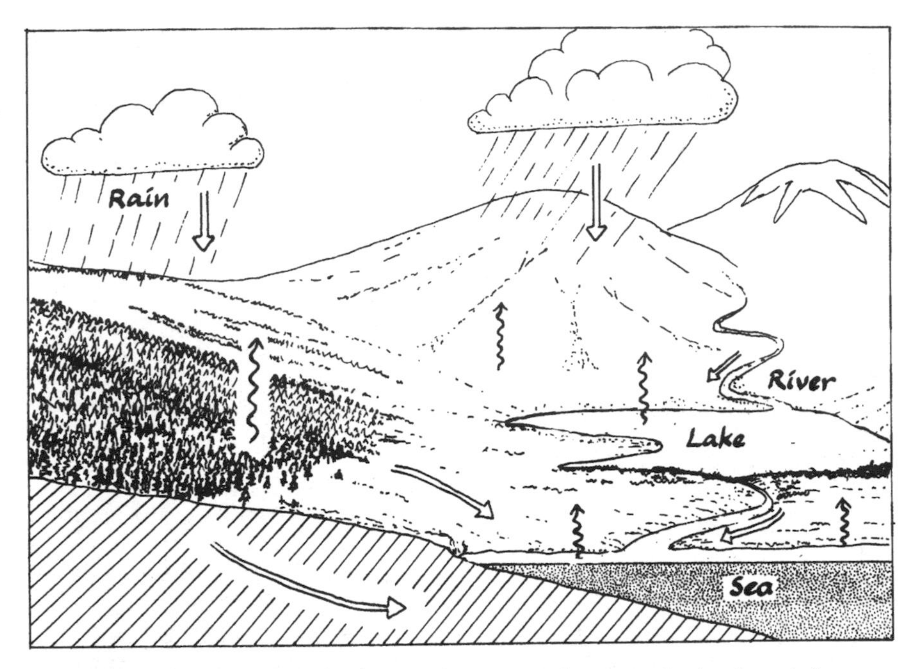

FIGURE 1.1. The water cycle. Fresh water falls, as rain and snow, on land and sea; it flows seaward, on the surface and below it (hollow arrows). Water vapor rises skyward from land and sea (wavy arrows), condenses into clouds, then falls again as rain and snow.

be about 1,120 kilometers long, approximately twice the length of Lake Superior.

The amount of *fresh* water in the world today is approximately 36 million cubic kilometers, a mere 2.6 percent of the total; of this fresh water, only 11 million cubic kilometers[2] (0.77 percent of the total, or 30 percent of the fresh water) counts as part of the water cycle, in the sense that it circulates comparatively fast. Of the water not in circulation, most is immobilized in long-lived polar ice sheets, and some is trapped, stagnant, under the ground.

No single answer can be given to the question, how fast does the water circulate? This is because the time taken by a given drop of water to complete the cycle, from ocean back to ocean, varies tremendously. It ranges from minutes or hours, as when a rainstorm blows inland from the sea, to thousands of years, the time during which a water drop may be frozen into a glacier. Indeed, there isn't a sharp distinction between circulating and noncirculating water: given enough time—hundreds of thousands, or millions, of years—all water circulates.

These statistics apply to what is happening at the present day. Until the 1980s, it was believed that the total amount of water on the earth had remained the same, with only negligible changes,[3] throughout the lifetime of the planet; in other words, that the earth's water cycle was virtually closed—no new water ever entered it and no existing water ever left it. New discoveries in the late 1980s have inspired new speculations. Many scientists now believe[4] that the earth's water supply is growing all the time, though not fast enough to solve humanity's water-shortage problems. According to the new theory, loosely packed "snowballs" of nearly pure snow, the size of small houses, are entering the earth's gravity field from the outer parts of the solar system every few seconds; they have been dubbed *small comets,* and most of them appear to weigh between 20 and 40 tons.[5] They melt and vaporize when they get near the earth. If this "rain of snowballs" has been going on at the current rate for billions of years (and there is no reason to suppose that it hasn't), the amount of new water we are continually gaining is equivalent to about 6 millimeters of rainfall over the whole earth, or 3 trillion tons of water, every 10,000 years.

The amount of fresh water on earth, as a fraction of the sum total of all water, fresh and salt, varies from one geological epoch to another. So does speed of the water cycle. At times when the whole world was warm—no ice sheets anywhere—and shallow, inland seas were more extensive than they are now, the amount of liquid water in circulation would have been much greater than it is at present, and the ratio of salt water to fresh would undoubtedly have been different. Conversely, at the height of an ice age, liquid fresh water was comparatively scarce, because much was immobilized in ice sheets. At the same time, the low temperatures would have reduced evaporation rates and precipitation rates, slowing the whole cycle.

To return to the present, to the fresh water the world holds now. People in general are at last becoming aware that human demands for fresh water are beginning to outstrip the supply. Quite probably, fresh water will turn out to be the factor that limits population growth,[6] as limited it obviously must be.

This book is not about the water crisis, however. It is about the natural history of fresh water. This does *not* mean the natural history of life in fresh water, which is an altogether different topic. Fresh water has a "natural history" of its own: the water cycle can be thought of as a "life cycle." Water comes from the sea as a vapor, travels for a time as a liquid, sometimes lingers as ice, and then returns to the sea again. Fresh water "behaves" in a number of different ways: it moves over and through the ground, sometimes fast and sometimes slowly; it pauses in lakes and ponds; it freezes; it vaporizes; it creates habitats for a wide range of ecosystems; it shapes the land. In brief, it is active and powerful, indeed more active and more powerful than the living things whose lives depend on it. The water cycle shown in figure 1.1 is such a convenient summary of what water does, and is so well known, that few of those who see it pause to think about the details. They are the subject of this book.

The water cycle has no beginning and no end, making any starting point as good as any other. We start, in chapter 2, with groundwater. At any one time, about 22 percent of all the fresh water on earth exists as groundwater,[7] out of sight and, for most people, out of mind. It is there all the same, moving in mysterious ways, in stygian darkness. It bears thinking about.

2

Water below the Ground: Groundwater

2.1 *Water Young and Old*

Think of all the fresh water on earth—liquid water, that is, disregarding polar ice. Then think of all of it that you can see—the water in rivers, lakes, and ponds. Sixty times as much water as this lies out of sight below the ground.[1] Water is under your feet almost everywhere you go.

All the water below the ground is called, straightforwardly enough, *underground water.* It occupies a variety of different underground spaces, ranging from tiny pores in the soil and narrow cracks in the bedrock, to enormous limestone caverns. But not all of it is "groundwater." The word *groundwater* has a narrower meaning: it applies only to water that saturates the ground, filling all the available spaces. The ground containing groundwater is known as the *saturated zone,* and the water's upper surface is the *water table.* Unless the water table is at ground level—as it is in swamps and marshes—there is an *unsaturated zone* above the water table in which the pores and cracks are only partially water-filled; they contain some air as well. This unsaturated water is called *vadose water,* and is the topic of chapter 4. In this chapter we consider true groundwater; how it gets where it is and how it is held there, how it moves around, and how it emerges at the surface.

Although all the water below the water table counts as groundwater, it isn't all of one kind: there are three different types. By far the most abundant is *meteoric water;* this is the groundwater that circulates as part of the water cycle. (Its name is related to *meteorology*—nothing to do with meteors.) A small proportion of the water deep underground does not circulate, however, and this noncirculating groundwater is of two kinds.

One kind is *fossil water* (sometimes called *connate* water): it is water that became trapped in ancient sediments when they were laid down, and that remained trapped as the sediment hardened into sedimentary rock.

The other kind is *juvenile water:* it is water given off by subterranean molten rock (*magma*), the material known as lava if it emerges from a volcano, or that forms igneous rocks if it crystallizes deep underground. Water is often an ingredient of magma, and when it cools underground, the water separates from the remaining ingredients to form pockets of liquid inside a mass of hard, cold, crystalline rock. This means that it has never been a part of the water cycle, but has remained where it first appeared, formed from the raw material out of which the whole earth is constructed; it could be thought of as newborn, being water that has not yet seen the outside world. This is the reason for calling it juvenile. Some of it (not all) is young in the ordinary sense, having only recently become separated from the magma of which it was a part.

The noncycling forms of groundwater, that is, fossil water and juvenile water, won't remain out of circulation forever. Given enough time, many millions of years in some cases, the rocks enclosing these waters will become eroded and the trapped water will be released. But for now these waters are immobile; they are not part of the great worldwide water cycle. In the sections to follow, we consider groundwater that does circulate. Section 8.5 describes groundwater immobilized by freezing, as happens in arctic permafrost.

2.2 *Water in "Solid" Ground*

Ordinary, meteoric groundwater is water that has soaked into the ground from the surface, from precipitation (rain and melting snow) and from lakes and streams. There it remains, sometimes for long periods, before emerging at the surface again. At first thought it seems incredible that there can be enough space, in the "solid" ground underfoot, to hold all this water.

The necessary space is there, however, in many forms. The commonest spaces are those among the particles—sand grains and tiny pebbles—of loose, unconsolidated sand and gravel. Beds of this material, out of sight beneath the soil, are common. They are found wherever fast rivers carrying loads of coarse sediment once flowed. For example, as the great ice sheets that covered northern North America during the last ice age steadily melted away, huge volumes of water flowed from them. The water was always laden with pebbles, gravel, and sand, known as glacial outwash, that was deposited as the flow slowed down.

The same thing happens to this day, though on a smaller scale, wherever a sediment-laden river or stream emerges from a mountain valley onto relatively flat land, dropping its load as the current slows; the water usually spreads out fanwise, depositing the sediment in the form of a smooth, fan-shaped slope. Sediments are also dropped where a river slows on entering a lake or the sea; the deposited sediments are on a lake floor or the seafloor at first, but will be located inland at some future date, when the sea level falls or the land rises; such beds are sometimes thousands of meters thick.[2]

In lowland country almost any spot on the ground may overlie what was once the bed of a river that has since become buried by soil; if they are now below the water table, the gravels and sands of such a riverbed, and its sandbars, will be saturated with groundwater.

So much for unconsolidated sediments. Consolidated (or cemented) sediments, too, contain millions of minute water-holding pores. This is because the gaps among the original grains are often not totally plugged with cementing chemicals; also, parts of the original grains may become dissolved by percolating groundwater (which is never as pure as distilled water), either while consolidation is taking place or at any time afterwards. The result is that sandstone, for example, can be as porous as the loose sand from which it was formed.

Thus a proportion of the total volume of any sediment, loose or cemented, consists of empty space. Most crystalline rocks (igneous and metamorphic) are much more solid; a common exception is basalt, a form of solidified volcanic lava, which is sometimes full of tiny bubbles that make it very porous.

The proportion of empty space in a rock is known as its *porosity*. But note that porosity is not the same as *permeability*, which measures the ease with which water can flow through a material; this depends on the sizes of the individual cavities and the crevices linking them (see section 2.3).

Much of the water in a sample of water-saturated sediment or rock will drain from it if the sample is put in a suitable dry place. But some will remain, clinging to all solid surfaces; it is held there by the force of surface tension but for which water would drain instantly from any wet surface, leaving it dry as a duck's back (for more on surface tension, see section 4.4). The total volume of water in the saturated sample must therefore be thought of as consisting of water that can, and water that cannot, drain away.

The relative amount of these two kinds of water varies greatly from one kind of rock or sediment to another, even though their porosities may be the same. What happens depends on pore size; if the pores are large, the water in them will exist as drops too heavy for surface tension to hold, and it will drain away; but if the pores are small enough, the water in them will exist as thin films, too light to overcome the force of surface tension holding them in place: then the water will be firmly held.

For any rock (consolidated or not), the proportion of the water in it that can drain away and the proportion that can't are called, respectively, the *specific yield* and the *specific retention* of the rock. Specific yield and specific retention together add to porosity.

Figure 2.1 shows the porosities, specific retentions, and specific yields for

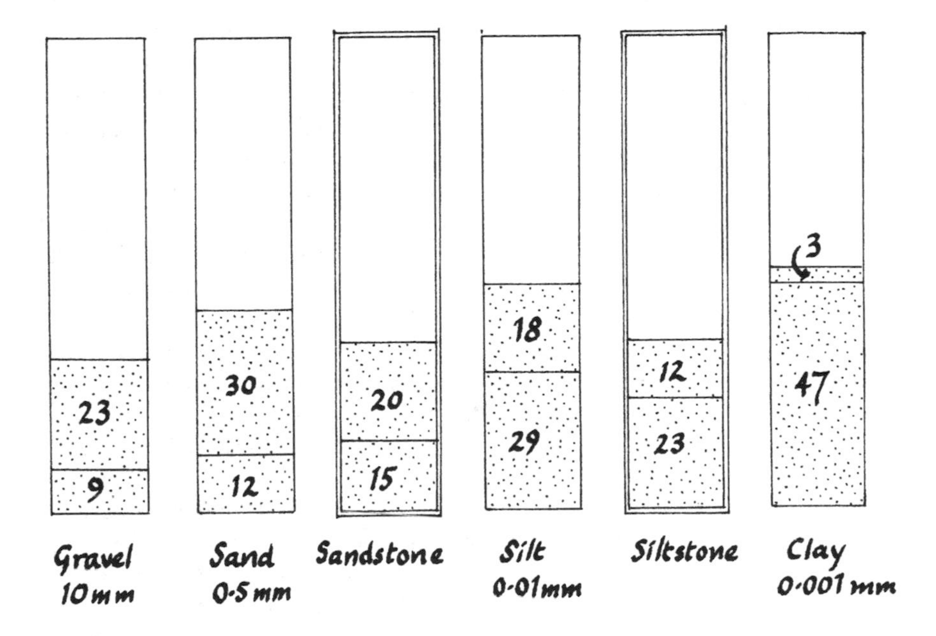

FIGURE 2.1. The porosities of some typical sediments and rocks. The average grain size (particle diameter) for each sediment is shown below its name. The stippled part of each bar shows the porosity of the material, made up of its specific yield (upper number) and specific retention (lower number), as percentages. Bars with single outlines, loose sediments; bars with double outlines, consolidated rocks.

typical samples of six different kinds of rocks[3] (it is convenient to use the word *rocks* for loose sediments as well as for rigid rocks). The results are averages; rocks of the same kind but from different places often differ widely in their characteristics.

Some generalizations are possible. For example, clay, with the finest particles, is much more porous than gravel, with the coarsest, which seems surprising until you consider the tremendous weight of wet clay. The material with the greatest yield is sand; although it is less porous than clay, it is much less retentive; gravel is still less retentive, but it yields less water than sand because it holds less—its porosity is less. As the figure also shows, the porosities of sandstone and siltstone are, respectively, less than that of sand and silt.

Sediments, whether loose or consolidated, are not to be found everywhere. The bedrock in many places consists of truly solid, crystalline rocks, wholly devoid of pores. How, then, can it hold groundwater? The answer is that "solid" rocks are nearly always fissured and fractured to some extent; also, rock now buried deep underground may have been crumbled by weathering when, in the distant past, it was a rock outcrop exposed to the atmosphere. Underground rock is not much more solid than that at the surface; such wholly undamaged blocks as do exist are never very large.

It is worth considering how rocks become split, fractured, and weathered. First, splitting: sedimentary rocks are laid down as "beds," that is, as layer upon layer of loose material that becomes consolidated later; it often happens that the beds are not strongly bonded, and in time splits develop between them. Some rocks can also split at an angle to their bedding plane; this is true of metamorphic rocks such as slate. Metamorphic rocks, whatever their origin, have at some time in their history been deeply buried under overlying rocks, so deeply that the tremendous heat and pressure has changed (metamorphosed) them. They become compressed and distorted, and liable to split along newly formed crystal faces.

Next, fracturing: there are many causes. Fracturing can happen below ground, as well as at the surface. Igneous rocks, formed from solidified magma, contract and crack as they cool. The gradual folding of rock layers that follows the pushing together of tectonic plates leads to the development of cracks where rocks are "stretched" over the top of a fold. The same thing happens when rock layers bulge upward because some overlying weight has been removed; the weight may be overlying rock, gradually eroded away over millions of years; or it may be an overlying ice sheet, formed during an ice age and then melted away over some thousands of years. Collectively, these fractures are *joints*. When the stresses causing rock to fold or bulge become so great that the blocks on either side of a fracture shift with respect to each other, the result is a *fault* or, more often, a whole series of little faults that together constitute a *fracture zone*.

Most of this splitting and fracturing occurs underground, where the cracks and crevices formed are immediately available as containers for groundwater. That there are an abundance of cracks and fractures in rock becomes evident when buried rock eventually appears at the surface, uncovered by erosion. Moreover, as soon as rock becomes exposed to the air, its existing splits, cracks, and fissures become enlarged by weathering. Weathering, like fracturing, takes place in a multitude of ways.

The commonest cause of weathering in cold climates is the freeze-thaw cycle: rainwater dribbles into a crack and subsequently freezes; as it freezes it expands, forcing the crack to open a little wider. After several repetitions of this process, the rock is likely to break; this is one of the ways in which quantities of rock fragments, ranging in size from boulders to sand grains, accumulate on the surface in mountainous country. In frost-free climates, cracks widen and rocks break because the rock itself alternately expands and contracts with the daily rise and fall of the temperature.

These expansions and contractions cause so-called mechanical weathering. Another form of weathering is chemical weathering. Rainwater, even in regions free of air pollution, is always weakly acidic; this is because it dissolves some of the carbon dioxide in the atmosphere as it falls, which converts it into dilute carbonic acid. Many types of rock are slowly dissolved by this acid. Limestone, in particular, dissolves with spectacular results: its cracks and fissures can become enlarged into enormous subterranean caves (see section 3.1).

Rocks of many kinds, including igneous ones such as granite, gradually crumble because some of the crystals of which they are composed are soluble in weak acid; as these crystals dissolve, the surviving crystals fall apart.

The final result of all this weathering is the production of the layer of broken, crumbled rock fragments that accumulates on top of exposed bedrock wherever the slope is too gentle for all the products of weathering to roll off or be washed away. This layer cannot hold groundwater when it is first formed because it lies above the water table. But in the geological future, perhaps millions of years hence, it will almost certainly become buried beneath newly formed sediments and soil; then it will be able to hold groundwater. Likewise, much present-day groundwater is held in ancient, long-buried layers of loose rock, lying where it was originally formed. So the "solid" ground beneath us is, indeed, riddled with holes. They range from pores, cavities, crevices, and cracks to spacious caverns and underground tunnels.

How deep is the deepest groundwater? Is there a lower limit to its occurrence? The answer to the second question is yes. Borings show that, at greater and greater depths, the spaces in rock become smaller and smaller. Because of the ever-increasing weight of the overlying material, and the steadily rising temperature, rock at great depths loses its rigidity; it becomes deformed under the tremendous pressure, until all pores and cavities finally disappear. This is believed to happen about 15 or 16 kilometers below the surface.[4]

2.3 *Moving Groundwater*

Groundwater is nearly always moving except, perhaps, at great depths. The rate at which it flows varies enormously but is "glacially" slow by surface standards; 30 meters per year is a typical figure. Ten to 15 meters per day counts as very rapid.

Two factors determine the rate at which groundwater flows. First is the nature of the ground itself. It must be *permeable* as well as porous; its pores and cavities must be linked by channels through which water can flow. A rock can be porous without being permeable; pumice, which contains a multitude of tiny bubbles, is an example; the bubbles are isolated from each other, making the rock impermeable even though it is buoyant enough to float on water. The permeability that allows groundwater to flow arises partly from the innate permeability of the buried rock, whatever it happens to be, and partly from the degree of fracturing and fissuring of the rock, as described in section 2.2.

The second factor determining the rate at which groundwater flows is the strength of the force, or pressure, pushing it. This depends on the *hydraulic head,* or simply the *head.* For example, imagine gently sloping ground with the soil overlying a thick layer of unconsolidated sediments; the sediments hold groundwater (figure 2.2). Imagine that the water table slopes in the same direction as the ground surface above it but much more gently, as would usually be the case. A borehole drilled into the ground anywhere on the slope begins to hold water as soon as it reaches the water table; the height of this water surface above mean sea level is the head.

To judge how the groundwater is moving, you need to compare the heads at two nearby sites. The difference between the heads is the *head loss* between

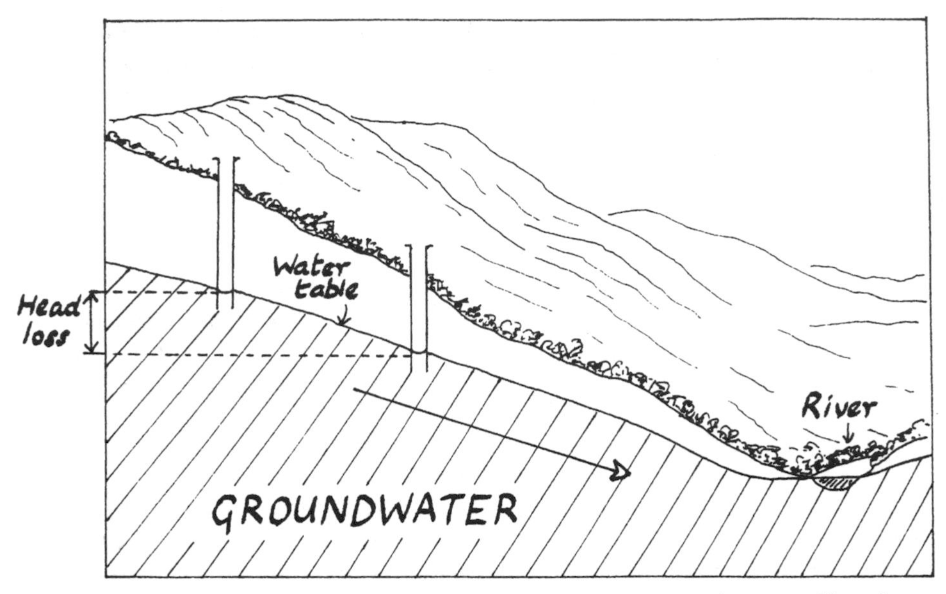

FIGURE 2.2. Two boreholes driven down until they just penetrate the water table and enter water-saturated ground. The difference in water level at the two borehole sites is the head loss between them. The arrow shows the direction of flow of the groundwater.

the two sites. Wherever there is a head loss, the groundwater must be flowing toward the site with the lower head. In the example, assume that one borehole is directly downhill of the other; then the head loss divided by the horizontal distance between them is the slope of the water table, known as the *hydraulic gradient.*

The rate at which the groundwater flows between the boreholes' sites is found by multiplying the hydraulic gradient by the *hydraulic conductivity* of the material through which the water is flowing. The hydraulic conductivity is the rate (in meters per day) at which water would flow through the material *if* the hydraulic gradient were unity, that is, *if* the water table sloped at an angle of 45°.This is a far steeper gradient than would ever be found in nature; hydraulic conductivities are merely numbers allowing the conductivities of different kinds of rocks and sediments to be compared. Hydraulic conductivity depends on other things besides the nature of the rock. It is also affected by the viscosity and density of the water flowing through it,[5] which depend, in turn, on the water's temperature and purity. Variations in the temperature and purity of groundwater are seldom large enough to make a noticeable difference, however. An exception is the rate of flow of industrial wastewater when it's hot; water at 40°C is less than half as viscous as water just above freezing point.

For the great majority of purposes, it is true to say that[6]

Rate of movement =
Hydraulic gradient × *Hydraulic conductivity*.

This is the rate of movement of the water as a whole, through the porous ground, and is technically known as the *specific discharge*. It is less than the velocity of the separate tiny "streamlets" flowing through each separate pore.

To consider some numbers, suppose two boreholes are 100 meters apart and the water level in the upper one is 2 meters above that in the lower one. The head loss is 2 meters. Therefore, the hydraulic gradient is 0.02, which means that the water table makes an angle of about 1° with the horizontal. If the groundwater is flowing through fine sand with a hydraulic conductivity of 1 meter per day, its rate will be 0.02 × 1 meter/day, that is, 2 centimeters per day.

How fast would the groundwater flow given the same hydraulic gradient but different earth materials? If the material were fine clay, which might have a hydraulic conductivity of 0.1 millimeters per day, the rate would be 2 micrometers (millionths of a meter) per day. If it were gravel with a hydraulic conductivity of 1,000 meters per day, the rate would be 20 meters per day.

Another way of looking at these flow rates is to consider the time it would take for groundwater to move a distance of 100 meters, again assuming a slope of 1°. The times for the clay, the sand, and the gravel are, respectively, 137,000 years, 13.7 years, and 5 days.

Because speeds of flow vary so enormously, the ages of different bodies of groundwater also vary enormously. The age is the time since the water left the surface and became part of the groundwater. Geologists can estimate the age of underground water using radiocarbon dating.[7] It has been found[8] that the water trapped in some Egyptian sandstones is up to 40,000 years old; it was old when the Pyramids were being built. Romanian limestones contain water between 15,000 and 25,000 years old. The deserts of North Africa and the Arabian peninsula overlie groundwater more than 35,000 years old, which presumably dates from wet periods during the last ice age. And some groundwater in the Columbia River basalts of Idaho, Oregon, and Washington is 32,000 years old.[9]

These groundwaters are truly old. Young groundwater is much more common, but judging its exact age is difficult; radiocarbon dating is too inexact. A judgment that can be made, however, is whether a body of water has been below ground since before 1953. The preceding year, 1952, was the year in which above-ground H-bomb testing began, and by 1953 the atmosphere and the rain had become heavily contaminated with tritium, a by-product of the explosions. Tritium is a radioactive isotope of hydrogen with a half-life of 12.4 years; although it decays quite rapidly, and although bomb testing stopped in 1967, the amount of tritium present in groundwater that originated in 1952 or since is still considerable. The amount in water dating from before then is negligible. The difference makes the two age classes easily distinguishable.[10]

2.4 *Aquifers*

Many different kinds of sediments and rocks go to make up the ground below the soil, and they are arranged, for the most part, in layers; the layers may be large or small, thick or thin, and can be of varying thickness and tilted at any angle. Intruding into these layers from below are occasional blocks of igneous rock formed of magma that has welled up from a greater depth into the overlying layered rock and congealed. Each of these different ingredients has its own characteristics so far as water is concerned: in particular, its own specific yield (section 2.2) and its own hydraulic conductivity (section 2.3).

Some layers may be aquifers. An *aquifer* is a body of rock or sediment that holds water in "useful" amounts; that is, the water is abundant enough, and can flow through the ground fast enough, for the aquifer to serve as a natural underground reservoir. Aquifers vary enormously in size and capacity. There are "good" aquifers and "poor" aquifers. What would be regarded as a "non-aquifer" in country with many good, easily accessible aquifers might count as a useful aquifer in a place where only small quantities of water were needed and there was no other source. It is all a matter of degree. Rock or sediment having a specific yield of 15 percent or more and a hydraulic conductivity of 10 meters per day would form a good aquifer anywhere, if there were enough of it. In any case, what defines an aquifer is its usefulness to humankind; the term has no precise scientific meaning.

For a layer to function as an aquifer, it must be large enough to store a useful volume of water and permeable enough for this water to be extracted, via wells and boreholes, at an acceptable rate. This last requirement means that to be useful for long periods the aquifer must be replenished by water coming in from elsewhere as fast as it is depleted. In a word, the aquifer must be constantly recharged (see section 2.5).

Because of the way sediments and rock strata are layered, it often happens that an aquifer is overlain by a *confining layer,* through which the water in the aquifer cannot escape upward; the confining layer might consist of almost impermeable clay, for example. Or the overlying material may form a so-called *leaky confining layer,* through which water from the aquifer can flow, though much more slowly than through the aquifer itself.[11] Leaky confining layers vary considerably in their leakiness, of course.

An aquifer overlain by a confining layer is a *confined aquifer,* and the water in it behaves differently from that in an *unconfined aquifer* (also called a *water-table aquifer*). An unconfined aquifer is saturated with water up to the level of the water table, and if an open-ended pipe is driven down into it, water will rise in the pipe until the levels coincide inside and out (see section 2.3 and figure 2.2). But if the pipe is driven into an aquifer below a confining layer, water rises in the pipe to a level higher than the top of the aquifer because it is under pressure.

As before, the height of the water in the pipe is described as a head, but now the meaning of the word *head* needs to be pondered. It is less simple than it appears: it has two meanings and implies more than it seems to. It is defined as the *height* of a water surface, which is simply a distance measured in a vertical direction. But it also, implicitly, means *pressure.* In the minds of hydrologists, the notions of pressure and of height are interchangeable, and we must consider why this should be.

Suppose you constructed a narrow pipe closed at the bottom by a hinged flap that would open inward when you wanted it to, but not before. Imagine the pipe driven deep into the groundwater. The groundwater exerts pressure on the hinged flap, and if you now open the flap, water will gush up the pipe. When it has come to rest and all is tranquil, visualize the motionless water at the bottom of the pipe, down by the hinged flap; it is *still* under pressure, but now there are equal pressures acting on it both from above and below, which cancel each other out. The upward pressure is the pressure that drove the water up the pipe in the first place: this pressure is still being exerted. The downward pressure is that due to the weight of the water in the pipe. Obviously the two pressures balance each other; if they didn't, the water level in the pipe would rise or fall until they did. The weight of the column of water in the pipe depends on its cross-sectional area as well as its height, of course. The cross-sectional area can be disregarded, however; it makes no difference, because whatever the area, the water in the pipe is pushing against exactly the same area of groundwater. The height alone suffices to measure the pressure.

This, in a nutshell, is why hydrologists use pressure and head (or height) as interchangeable terms. Meteorologists do the same thing, when they use the height of a column of mercury in a barometer as a measure of atmospheric pressure.

A pipe used to find the depth of the water table, or the head at any depth, is called a *piezometer;*[12] the derivation is from the Greek *piezo,* pressure. The amount of water that an unconfined aquifer can hold in storage is found simply by multiplying the saturated volume of the aquifer by the specific yield (see section 2.2) of the aquifer material. The amount that can be held in a confined aquifer is slightly affected by the pressure it is under; the pressure compresses the aquifer material and reduces its porosity; the effect of compression only becomes appreciable in aquifers at considerable depth, however.

Now imagine piezometers driven down through a confining layer into a confined or partly confined aquifer (figure 2.3). The water will rise to different levels in them, levels that correspond with an imaginary surface known as the *piezometric surface.*[13] The piezometric surface is related to the water in a confined aquifer in the same way that the water table is related to the water in an *un*confined aquifer. In both cases it is the surface to which water rises when a pipe is driven into the aquifer. The only difference is that, whereas a water table is a "real," tangible surface, a piezometric surface is, in a sense, "imaginary." It is where the water level *would* rise to *if* the pressure exerted on it by the confining layer were removed. The piezometric surface is sometimes below and sometimes above ground level; in either case the aquifer, and the water in it, are described as *artesian,* and a well drilled into it is an *artesian well.* Often, however, the term *artesian well* is used only when the piezometric surface lies above-ground at the site of the well, so that water gushes from it spontaneously, without pumping. The word derives from a place-name: Artesium in Roman Gaul, which is now the province of Artois, in northeast France.

The difference of the water levels in two piezometers is the head loss between them, and the head loss divided by the distance separating them is the hydraulic gradient, as before. Given a gradient, water will flow, but to measure the flow rate through an aquifer, the notions of hydraulic conductivity

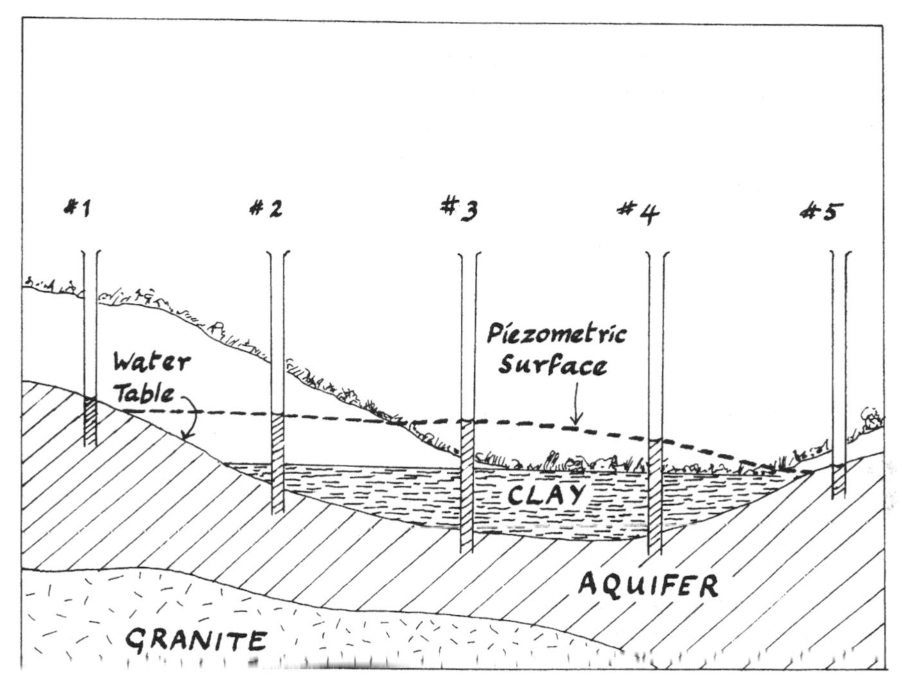

FIGURE 2.3. A row of five piezometers driven down into a partly confined aquifer. The confining layer is the clay of an old lake bed, partly buried under more recent deposits of permeable sediments. Piezometers #1 and #5 penetrate unconfined parts of the aquifer, and encounter water when they reach the water table. Piezometers #2, #3, and #4 go through the confining layer of clay before reaching the aquifer; water rises into them to the level of the piezometric surface, which is below ground in #2 and above ground in #3 and #4.

that we considered above are not particularly useful as they may well be different at different levels in the aquifer. It is better to consider the volume of water passing, per second, through a vertical cross-section of the whole aquifer. This depends on the average hydraulic gradient, the width of the aquifer, and a quantity called the aquifer's *transmissivity,* thus:[14]

> *Volume of water per second =*
> *Transmissivity × Aquifer width × Hydraulic gradient.*

Transmissivity is measured in square meters per second. It depends on both the thickness and the permeability of the aquifer.

The water moving within an aquifer must obviously come from some source and go to some destination. These are the topics we turn to next.

2.5 *The Recharge and Discharge of Aquifers*

The water in any aquifer circulates as part of the water cycle. An aquifer gaining water becomes *recharged,* and an aquifer losing water *discharges* it. How do these processes happen?

The most obvious and important source of new water is falling rain and melting snow. Not all this water sinks into the ground and then down as far as the water table to become groundwater, however. Part of it evaporates from the surface, part is absorbed by plants as it percolates down through the soil; and if the water arrives too fast for all of it to soak in, some flows away over the surface. But with heavy rain or copious snowmelt, some water will soak in and recharge the groundwater. The water will percolate straight down if an unconfined aquifer lies directly below; but if a confining layer blocks downward movement, the water will be deflected, to recharge an aquifer some distance away (see section 2.7).

It often takes considerable time for rainwater to flow through the unsaturated zone on its way to the water table. The time depends on the distance the water must go (the thickness of the unsaturated zone), on the permeability of the unsaturated zone, and on the amount of water already in this zone, which affects its permeability: surprisingly, the drier the soil, the slower the flow. In desert country it may take years for the water supplied by a rainstorm to recharge the groundwater.[15]

Groundwater is also recharged to some extent from surface water—rivers, lakes, and wetlands. Many lakes leak, though not necessarily through the whole lake floor: leaks are often localized. The same thing happens with rivers flowing through permeable material. A river's surface is usually a narrow strip of the water table exposed above ground. If the water table slopes down from either side of the water surface in the channel, some of the river water will soak into the ground and recharge the groundwater (see figure 2.4a). This happens when a river is in flood but seldom at other times.

Precipitation, and leakage from natural wetlands, lakes, and rivers are the processes that recharge groundwater in unoccupied country. In populous regions, some of an aquifer's recharge comes through human agency, both inadvertent and deliberate. Water seeps from artificial reservoirs in the same way that it does from lakes; it also leaks from buried water pipes and sewers (see

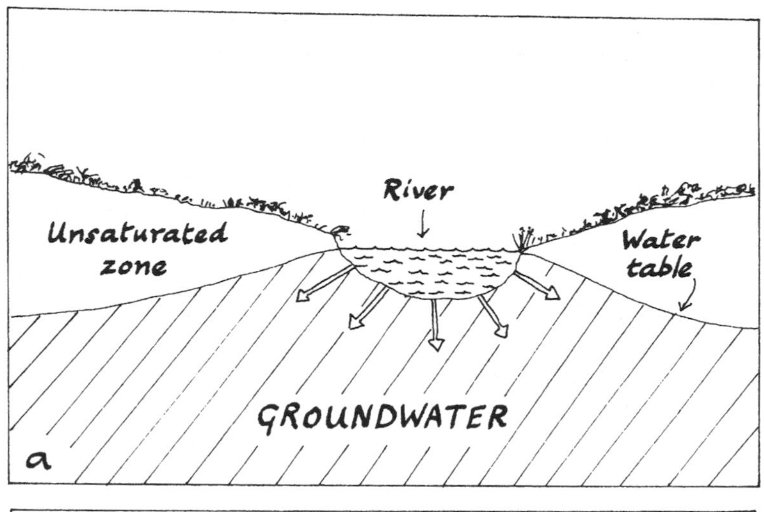

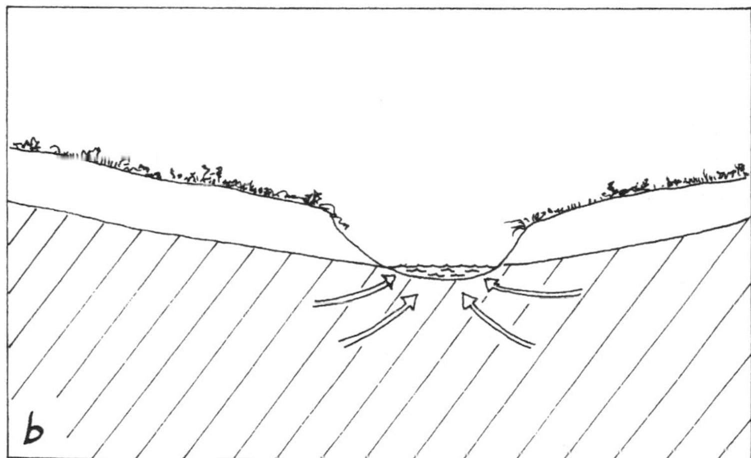

FIGURE 2.4. *(a)* Groundwater being recharged by a river in flood. *(b)* Groundwater discharging into a river at low flow. Note the slope of the water table in each case. The arrows show the direction of flow of the groundwater.

section 3.5). A proportion of the irrigation water distributed to farms from a reservoir soaks down into the groundwater; so does much of the water lavished on lawns, gardens, city parks, and golf courses. An aquifer is sometimes recharged artificially, by pumping water into it from some other, less heavily used aquifer.

In uninhabited regions the recharging of groundwater is balanced by its discharge; if this were not so, the water table would rise continuously, which obviously doesn't happen: disregarding seasonal fluctuations, water-table levels remain more or less constant over years and decades—though not, perhaps, over centuries. The only way that the ground can lose its water naturally is by seepage and springs.

Much groundwater seeps into lakes and rivers through their beds. When an isolated pond persists without drying up, after a prolonged period of hot, dry weather, the most likely explanation is that what the pond loses to evaporation is regained through submerged seeps or springs, in other words, by groundwater entering it invisibly from below. Likewise with a stream that keeps flowing during a dry period, with no visible source of water supplying it—no rain, no snowmelt, and no headwaters lake. Such a stream is evidently receiving groundwater discharge somehow: figure 2.4b shows what is probably happening.

At the coast a certain amount of groundwater seeps into the sea. On land, groundwater emerges as a seep or spring wherever the water table reaches the surface. There is no clear distinction between seeps and springs. Water oozes slowly from a seep and flows briskly from a spring, but often it dribbles from the ground at a rate that would make either term acceptable. Slow seeps are commoner than vigorous springs and are apt to go unnoticed; their presence can be suspected wherever you see groups of moisture-loving plants growing in an otherwise dry area, and especially if the groups are aligned along a contour across sloping ground (figure 2.5), forming a *spring-line*. For the most part, however, the discharge of groundwater goes unnoticed.

A stream of water emerging from the ground isn't always a true spring, fed by groundwater. Underground streams also flow through the unsaturated zone and emerge here and there at the surface. This often happens in mountainous country (see section 4.3).

Recognizing a true spring can be difficult. The dissolved chemicals in a true spring correspond with those in groundwater obtained from a nearby borehole. And, unless the spring comes from a very shallow aquifer, with the water table less than 3 meters down, the temperature will remain constant all through the year. The water will seem chilly in high summer and surpris-

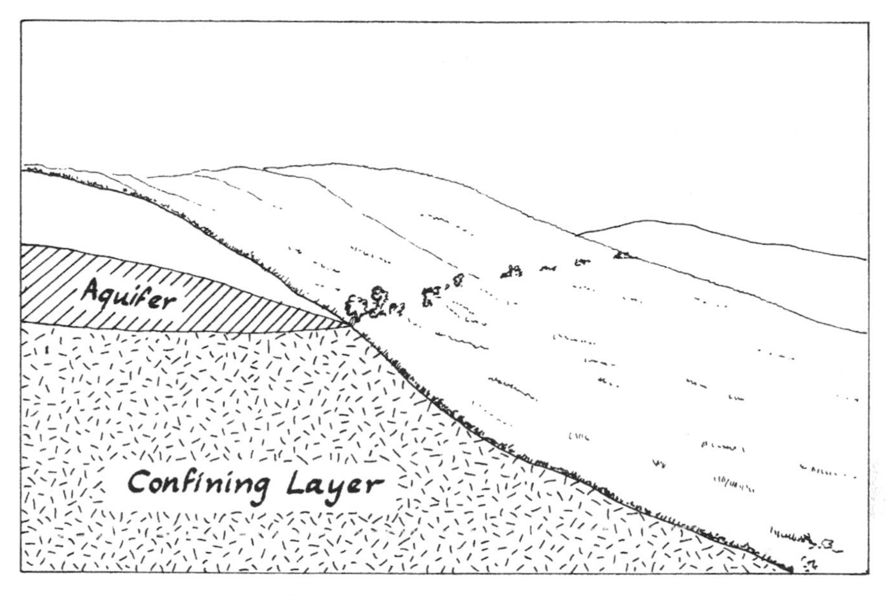

FIGURE 2.5. A spring-line, marked by clumps of shrubs, following a contour across a dry hillside. The groundwater is in an unconfined aquifer above a confining layer.

ingly warm in the depths of winter. In a cold climate a spring will continue to flow all through winter, even when neighboring ponds and streams are frozen solid.

Aquifers are discharged as a result of human intervention wherever water is pumped from a well or borehole for domestic use, for irrigation, to meet the requirements of industry, or to top up a depleted aquifer somewhere else. Natural groundwater is also abstracted wherever mines or quarries are "dewatered" (pumped dry) to prevent flooding. For more on the artificial depletion of aquifers, see section 2.7.

2.6 *Which Way Does Groundwater Flow?*

Groundwater can flow in any direction. It moves in response to hydraulic head, that is, pressure (see section 2.3). It is impossible to *see* which way groundwater is flowing, of course, because it is hidden. The only way to infer what is happening out of sight below ground is to find out the direction in which the pressure is acting. This is done with piezometers (section 2.4). To find out whether the pressure, and therefore the direction of flow, below one particular spot on the ground is upward or downward, a *nest of piezometers* is used.

This is a group of several piezometers all in a single borehole, with their open (buried) ends at different depths (figure 2.6). If the level to which the water rises is least in the piezometer whose open end penetrates deepest, then the pressure is least at the bottom of the nest and the groundwater flows downward; if the piezometers are probing an unconfined aquifer, this means that the nest is located in a recharge area, where water is sinking into the ground. Conversely, if the water in the deepest piezometer rises the highest, the pressure is upward, and the groundwater flows upward; if the piezometers are probing an unconfined aquifer, the nest is in a discharge area, where water is being forced out of the ground. If the water rises to exactly the same height in all the piezometers, the pressure, and likewise the flow (if any), must be precisely horizontal.

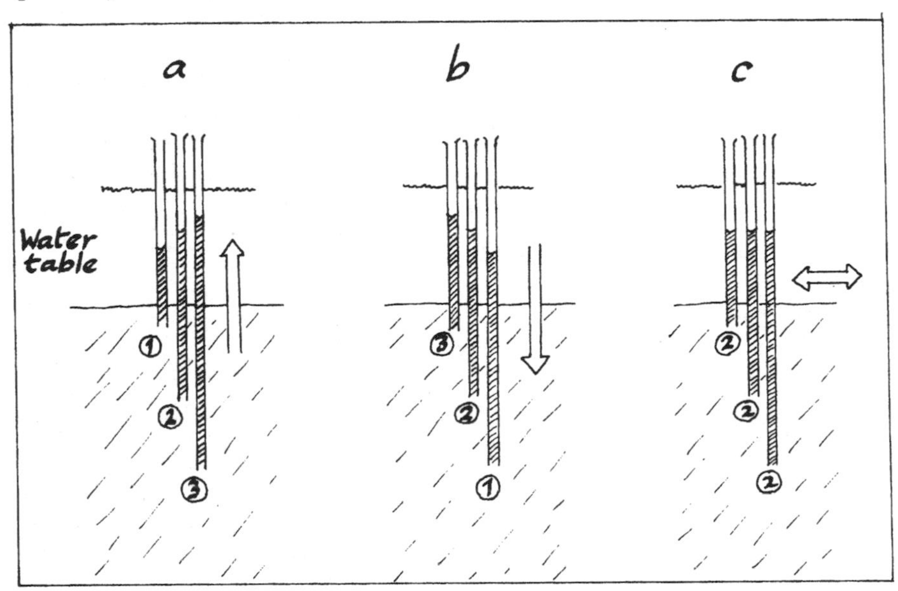

FIGURE 2.6. Three nests, each of three piezometers. For each piezometer, the hydraulic head it records is shown in the circle at its tip. The hollow arrows show the direction of flow of the groundwater. *(a)* Greatest head in the deepest piezometer: flow upward; *(b)* greatest head in the shallowest piezometer: flow downward; *(c)* head equal in all three piezometers: flow (if any) in an indeterminable horizontal direction.

To say groundwater is flowing *upward* does not necessarily mean that it is flowing straight up. It means merely that its direction trends upward, perhaps steeply, perhaps gently. And correspondingly with *downward.* To judge the exact direction of flow, there must be piezometers at many different sites, probing to many different depths. The height of the water in each piezometer is the hydraulic head at the tip of the piezometer.

Once the hydraulic head at many different depths at numerous sites is known, it becomes possible to conjure up a mental image, in three dimensions, of what is going on below ground—in the north/south, east/west, and up/down directions. This mental image cannot, however, be drawn on paper in the same way as an ordinary contour map of the surface of the land, showing hills and valleys, and rivers flowing down the valleys. There are two differences. First, groundwater is flowing through a solid, rather than over a surface, and may be flowing in different directions at different levels. Second, the water does not necessarily flow along narrow channels; in an aquifer of porous, permeable material, it may flow throughout the whole body of the aquifer.

These difficulties are overcome by drawing a cross-section of the ground. Figure 2.7a is a simplified example of such a cross-section. It assumes, for simplicity, that all the piezometers are lined up in a single east-west row. They are sampling an unconfined aquifer at various depths; the numbers show the heads recorded at numerous locations. (To keep the diagram clear, low numbers have been used; their magnitudes are arbitrary.) With this information it becomes possible to draw *equipotential lines* (shown dashed) on the diagram; each is drawn so that every point on it has the same hydraulic head (printed at the end of the line). These lines cannot be exact, of course; they are interpolated among the locations at which the heads were actually measured.

Now consider how the groundwater must be flowing given the pattern of equipotential lines shown in figure 2.7a. The flow is always from higher to lower heads, so the *flow lines* (arrows) are as shown in figure 2.7b; water flows down into the ground where the water table forms a "hill"; it flows vertically down from the "summit" and obliquely from the slopes. It flows horizontally where it crosses a vertical equipotential line, as that is where the hydraulic head is the same at all depths (as in figure 2.6c). In the valley the water flows upward. The whole pattern of curves representing equipotential lines (dashed) and flow lines (dotted arrows) is known as a *flow net.*

When examining the figure, keep in mind that it shows the vertical scale tremendously exaggerated relative to the horizontal scale. This is nearly al-

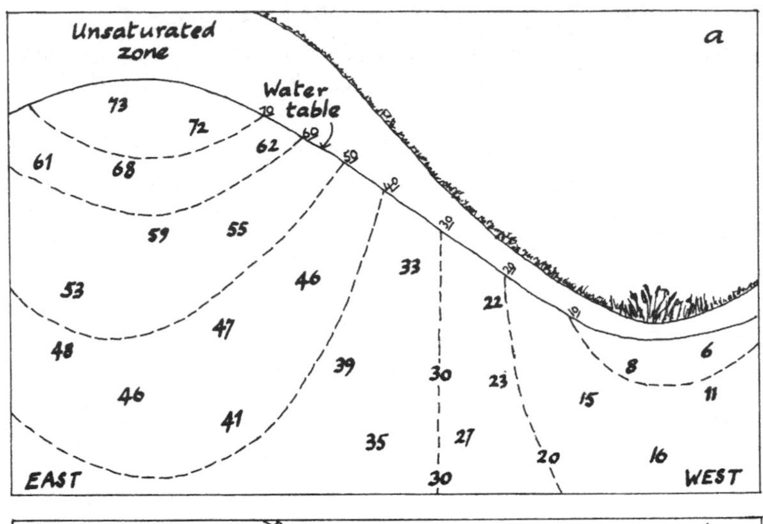

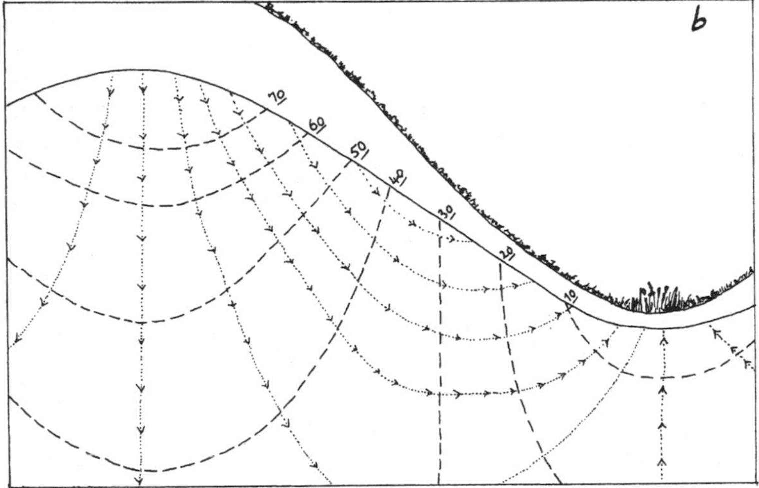

FIGURE 2.7. *(a)* A cross-section through the ground showing the hydraulic heads (numerals) at several points below the water table. The dashed lines are equipotential lines; the corresponding heads are shown (underlined) where the lines cross the water table. *(b)* The same section with the numerals deleted and groundwater flow lines (dotted arrows) drawn in.

ways done in hydrological diagrams, for clarity. If the scale were the same in both directions, gentle hills would appear indistinguishable from flat land. Therefore, slopes in the diagram are much steeper than the slopes they represent in real life.

The water table is an equilibrium surface at least for short periods of time—it rises and falls gradually, as wet weather alternates with dry. When rain soaks into the unsaturated zone, the incoming water flows downward into the groundwater from the top of the water table "hill" as fast as it arrives. Meanwhile, at the bottom of the water-table "valley," water is being sucked upward by unsaturated soil and subsoil, which are themselves depleted as plants extract water with their roots. The water table maintains its position only for as long as the input and output balance.

2.7 *The Anatomy of the Subterranean World*

The invisible, three-dimensional world below the ground is as diverse and complex as the visible landscape at the surface. Admittedly, it is in total, perpetual darkness. It is solid almost everywhere, and inaccessible to everybody except miners, tunnel makers, and cave explorers; even they see only a few selected parts of it. At levels below the water table, it is completely saturated with water. And, as the depth increases, the temperature rises steadily. This runs counter to intuition for most people: dark and wet surely suggest cold. In fact, the temperature rises at between 1° and 3°C per 100 meters.[16] It is a scene difficult to visualize, but the effort is worth making.

The most important fact about this dark, wet world from the hydrological point of view is that it is far from uniform. As we saw in section 2.4, it usually has a layered structure, with the layers differing from each other in hydraulic conductivity: several aquifers and several confining layers, each with its own geological characteristics, are often stacked up layer upon layer, and each separate confined aquifer has its own, individual piezometric surface.

We have so far considered the way water enters, moves through, and flows out of small parts of individual aquifers. Now we fit the parts together, and consider water movements in large, relatively complicated "solid landscapes" below the ground.

To begin, consider a single, unconfined aquifer, lying under a large tract of highland country with a multitude of hills and hollows (figure 2.8a). The surface of the water table has corresponding hills and hollows, though with much gentler slopes (the slope of a water table seldom exceeds 2°, a fact to

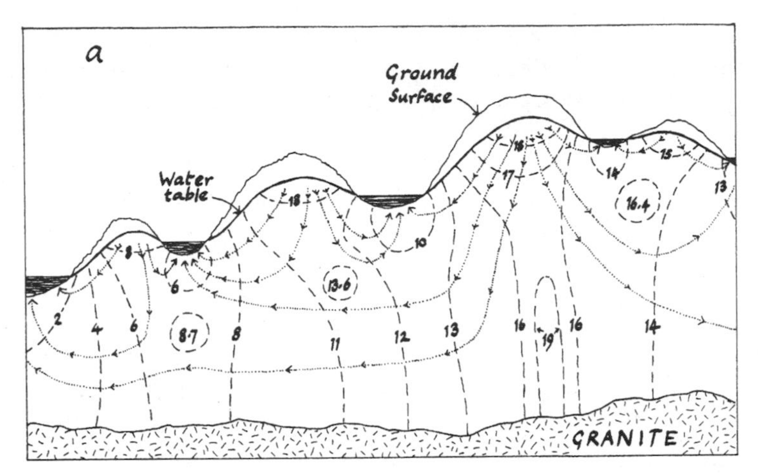

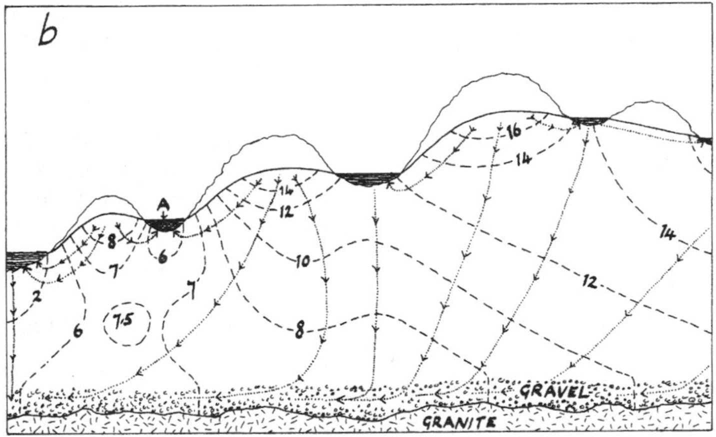

FIGURE 2.8. *(a)* A section through hilly country with five small lakes (the hills appear unnaturally steep because of the exaggerated vertical scale). The dashed lines are equipotential lines; equipotential lines around stagnation zones form closed loops. The numbers show hydraulic heads. The dotted arrows are flow lines. *(b)* A section with a layer of permeable gravel just above the confining granite. The water-table undulations are gentler than in *(a)* even though the surface topography is the same. Only the lake marked *A* is "leak-proof" (with a stagnation zone below it). In the other lakes, water seeps in near the shores and out at the center. On the right, water seeps from one lake to another. Much downward flowing water reaches the permeable gravel layer and flows away horizontally.

keep in mind when looking at diagrams like this one, with greatly exaggerated vertical scales). The climate is wet, and in most of the hollows the water table emerges above the ground, creating little lakes. You could say that the lakes are groundwater "outcrops."

At first thought, one might suppose that each little lake is fed only by water from the hills immediately surrounding it, and for some of the lakes this may be true. But other lakes may receive only a fraction of their water from the nearest hills; the rest comes from more distant hills, as can be seen from the flow net in the figure. Every hill, however small, serves as a groundwater recharge area: that is, some of the rain falling on it will soak down and become added to the groundwater, even though in heavy storms some will remain on the surface and flow directly into the nearest lake. The destination of water soaking into the ground depends on exactly where it reaches the water table, and on the pattern of equipotential lines below the water table.

As figure 2.8a shows, some flow lines follow a shallow path and emerge at the lake nearest them; others penetrate deeper and return to the surface farther away. As a result some of the lakes are fed by water that has traveled underground for only a short distance, others by water that has traveled a long way underground. This makes a difference to water chemistry. Water that has flowed a long way through rock, and been in contact with it for a comparatively long time, will contain more dissolved minerals than water that has made a short, quick journey. The contrast may be revealed by the plants growing in the shallows around each lake. For example, in a landscape in Saskatchewan, where the ground is hummocky because of moraines left by the ice sheets of the last ice age, the numerous lakes are not all the same, either chemically or botanically.[17] The grasses and sedges growing around lakes with fairly pure water, from nearby high ground, differ noticeably from the bulrushes and cattails that grow in lakes richer in nutrients, whose water has traveled from distant recharge areas. The most highly mineralized lakes may have only sparsely vegetated shores, with scattered clumps of a variety of rush peculiar to saline or alkaline soils.

Figure 2.8a also shows three *stagnation zones.* The water is almost motionless in each zone because, at some point within it, the hydraulic head is greater than at any other point in its neighborhood. Thus all flow lines lead away from the point: every drop of water flowing toward it is slowed and deflected. Except for imperceptible leakage, the water in a stagnation zone remains where it is, without being replaced, until (as is bound to happen sooner or later) the geometry of the flow net changes. A lake underlain by a big enough stagnation zone cannot leak. All seepage will be upward, feeding groundwater into the bottom of the lake. Upward seepage in such a lake can easily be detected, using a simple device made of a tube and a plastic bag, in which the water collects.[18] Another point to notice is that water seeping up through a lake bed close to its shores has made a shorter, shallower underground journey than has water entering near the lake's center. Because of this, seeps near the center contain a higher concentration of dissolved minerals than do near-shore seeps.[19] The difference can be as pronounced as that between separate small lakes, already described.

Water-table lakes do not always have stagnation zones below them. Figure 2.8b shows a landscape in which only one lake is underlain by a stagnation zone, making it almost "leak-proof"; in each of the other lakes, water seeps down through the lake bed and recharges the groundwater. The contrast between the two landscapes results from what lies beneath the water-table aquifer:[20] in figure 2.8a the aquifer rests on a confining layer of impermeable rock. In figure 2.8b a thin layer of highly permeable gravel is sandwiched between aquifer and bedrock; water can flow rapidly through the gravel, and it causes the equipotential lines, and hence the whole flow net, to take on an altered pattern. One of the changes is the flattening of the water-table "hill" between the two right-hand lakes; this allows water to leak from one lake into the other.[21]

Both parts of figure 2.8 show an unconfined aquifer. Now let's consider how groundwater flows when the ground has a layered structure, so that the water in a deep aquifer is trapped beneath a confining layer. Figure 2.9 shows an example. It has an upper, water-table aquifer and a lower, confined aquifer of the same material, with the same hydraulic conductivity. They are separated by a horizontal, leaky confining layer (the layers are unshaded to allow the equipotential lines and flow lines to show clearly). Note how the flow lines behave when they cross the boundaries between the aquifers and the confining layer: on going from an aquifer into the confining layer, they bend toward the vertical, and vice versa when they emerge from the confining layer into one or other of the aquifers.[22] This is true of horizontally layered ground in general: water tends to flow *along* the permeable layers and *across* the confining layers; the rates of flow are vastly different, however, as we shall see.

In figure 2.9 some of the water enters and leaves the upper aquifer without ever reaching the confining layer. Some (only a little) enters the confining layer and flows within it before emerging into the upper aquifer again. And some (right half of the figure) leaks through the confining layer to recharge the

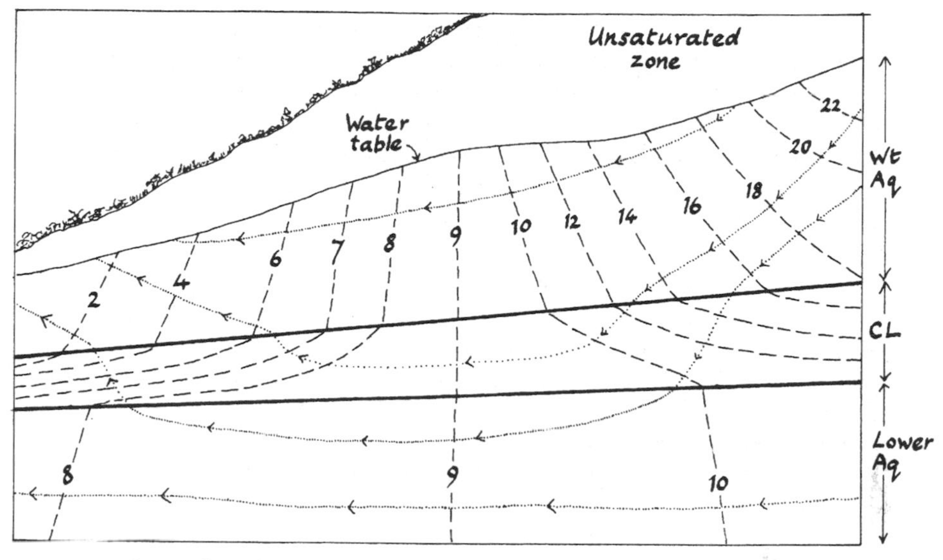

FIGURE 2.9. Section through a sloping landscape with two aquifers, an upper, water-table aquifer *(Wt Aq)* and a lower, confined aquifer, separated by a confining layer *(CL)*. The dashed lines are equipotential lines, labeled with their hydraulic heads. The dotted arrows are flow lines.

lower aquifer; water from the lower aquifer also leaks upward, back into the upper aquifer (left half of figure). Obviously, a well drilled down to the confined aquifer near the left edge of the figure, where the hydraulic head at the surface is less than 2, would turn out to be a spontaneously flowing artesian well (see section 2.4), as the head in the confined aquifer directly below is almost 8.

But note, as well, that the arrows show only the direction of flow; they reveal nothing about its speed or volume. The speed depends on the hydraulic gradient and the hydraulic conductivity of the material the water is flowing through (see section 2.3). The hydraulic gradient is much greater in the upper, unconfined aquifer than in the lower, confined one, as can be seen by comparing the spacing of the equipotential lines; water flows faster in the upper aquifer. In useful aquifers, the flow rate is ordinarily in the range 10 centimeters to 3 meters per day.[23] In the example, as is usual in nature, the hydraulic gradient is greatest of all in parts of the confining layer. But this does not mean the water in this layer flows fast: quite the contrary. Because of the extremely low hydraulic conductivity of the layer, water can barely move within it.

The mobility of water is much greater in the aquifers than in the confining layer because they are much more porous—that's what makes them aquifers. The displacement of whole bodies of water depends on both their volume and their speed of flow. Thus the displacement is less in the confined aquifer than in the unconfined one, because the water in the confined aquifer is moving more slowly: although its hydraulic conductivity is high, the hydraulic gradient is low.

The foregoing paragraphs do no more than hint at the structural complexities below ground. A complication worth noting is that many rocks are *anisotropic;* that is, their hydraulic conductivity is not the same in all directions. Typically, rocks and sediments conduct water far more readily in a horizontal than in a vertical direction. Unequal conductivity makes a big difference to the patterns of flow nets.[24]

It must be emphasized that the flow nets in figures 2.8 and 2.9, indeed all flow nets, show what is going on at only a single moment in time. Hydraulic heads are forever changing, and with them the pattern of the equipotential lines and the directions of the flow lines.

The speed at which things change varies widely. For example, rapid changes can happen when heavy rain follows a long dry spell. At first, because the ground is so dry, the water table may be a long way down: it may be far below the surface hollows and valleys where groundwater normally seeps out (discharges); then rain drains into the hollows, turning them into recharge, instead of discharge, areas. The water percolates down to the water table and forms a *groundwater mound* directly under each surface hollow before it has time to spread laterally. So for a short period the water-table topography is more or less a mirror image of the surface topography above it; below ground, water flows away from the mounds, in the opposite direction to its flow up at the surface.

A groundwater mound is a small-scale example of a more general phenomenon. As we noted at the beginning of this section, the undulations of the water table in hilly country usually match the visible undulations of the ground surface: but not always. When they fail to match, the direction of flow below ground differs from that above ground. The invisible "hills" and "ridges" of the water table below your feet are *groundwater divides,* in the same way that visible hills and ridges in the ordinary landscape are divides for surface water. The flow patterns on the surface often reveal nothing about what is going on below ground.

3

Groundwater in Use

3.1 Useful Aquifers of Different Kinds

Everyone should know the source of their water. Some people depend on groundwater pumped from aquifers, and some on surface water from lakes, reservoirs, and rivers. Treating North America as a whole, groundwater supplies only about 20 percent of the total water used, but this figure conceals tremendous variation from place to place: in the dry Southwest, in Texas and Colorado for instance, more than 45 percent of the water supply comes from groundwater, whereas in rainy New England the figure is less than 1 percent.[1] The percentage of the water used in family homes that consists of groundwater is much greater than these numbers suggest; this is because rural families who must make their own private arrangements for a water supply generally find groundwater cheaper, cleaner, and more reliable than surface water. In Canada, the proportion of the population relying on groundwater for household use seems to be rising: between the 1960s and the 1980s, it rose from 10 percent to over 25 percent.[2]

Although users of surface water are in the majority, they are not independent of groundwater. Water is forever changing from groundwater to surface water and back again, so that even those who use surface supplies get at least some water that has spent time below ground. So knowing the source of your water supply entails knowing something about the aquifers in your region.

Aquifers differ from each other in many ways: in the quantity of water they can hold, the ease with which it can be extracted, the purity of their water, and, above all, the geological material—the kind of rock—they are made of. It is this last point that matters most, as it explains all the other differences. And although climate controls the amount of water that goes into an aquifer, it does not control the way it comes out; that is why, as we shall see, a sandstone aquifer in Wisconsin, for example, resembles a sandstone aquifer on the Pacific coast more closely than either of them resembles an unconsolidated sand and gravel aquifer not far away from the sandstone one in Wisconsin. Indeed, entirely dissimilar aquifers can be stacked one atop another at the same location; thus, in much of peninsular Florida, an aquifer of sand near the surface lies above a limestone aquifer tens of millions of years older.

Here are examples of the kinds of geological structures that give aquifers their distinctiveness.

To begin with sandstone aquifers: figure 3.1 shows sections through the ground to show the "lay of the rock" in two typical examples.[3] Note the resemblance: figure 3.1a shows an aquifer near the tip of Lake Michigan, lying across the Illinois-Wisconsin border; figure 3.1b shows an aquifer in the Gulf Islands of the Strait of Georgia, the strait separating Vancouver Island from mainland Canada (see figure 3.2). A difference the figure doesn't show, which is immaterial from the water-holding point of view, is the difference in age of the sandstones in the two places; the sandstone in the Illinois-Wisconsin aquifer is more than 400 million years old, that in the Gulf Islands less than 100 million years old. But in each case tilted beds of sandstone are partly capped by a confining layer of shale and overlie another confining layer (crystalline rock in the American example, shale in the Canadian); the aquifers are recharged where the confining layers don't cover them. In both cases water is held in the cracks and fissures of fractured sandstone, and in both cases, the groundwater is partly artesian (see section 2.4). To the casual observer, the most obvious difference between the two aquifers is that the American one is more deeply buried than is the Canadian, which outcrops at the surface in many places. Note, too, the difference in the scales of the two diagrams.

Next, consider unconsolidated aquifers of sand and gravel in *glacial drift*. Glacial drift is all the material eroded by the huge ice sheets of past ice ages and left lying on the surface of the ground. It is far from uniform. Some of it consists of sands and gravels—good aquifer material—deposited by the wide rivers flowing from the ice when it melted; some of it is glacial *till*, which

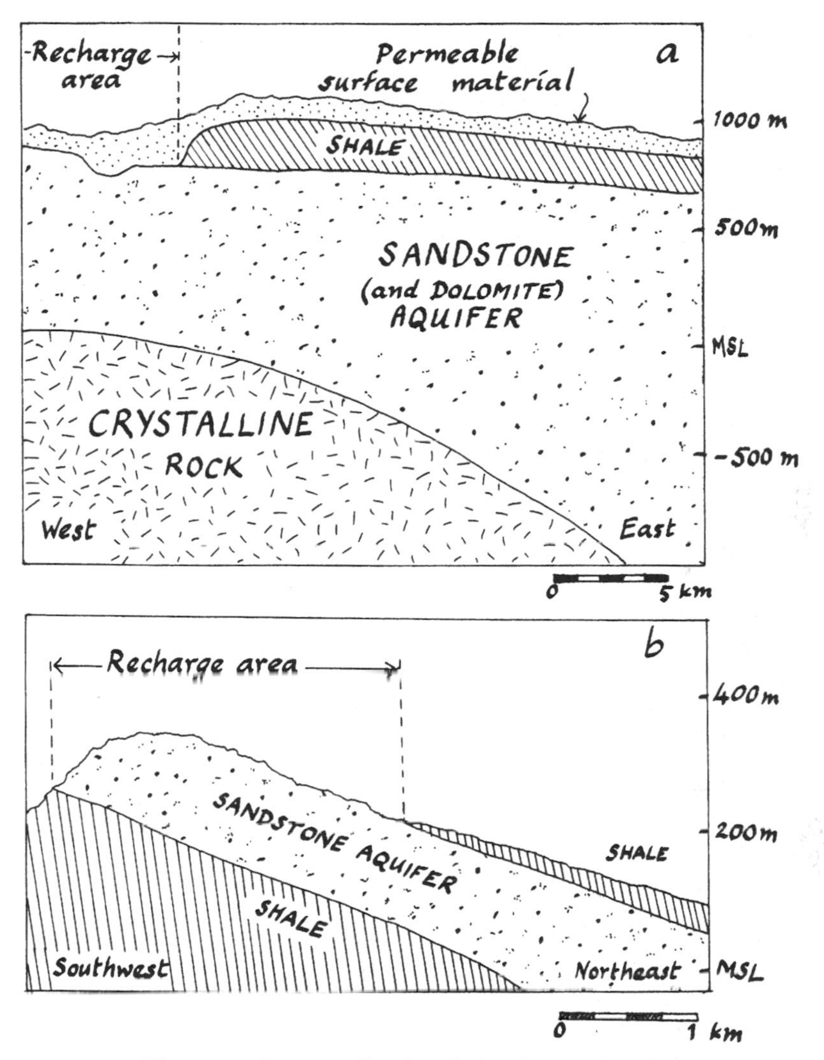

FIGURE 3.1. The two sandstone aquifers described in the text. Note the contrasting scales of the two diagrams; in both, the vertical scale gives the elevation in meters above and below mean sea level *(MSL)*. (*a*) On the Illinois-Wisconsin border, just west of Lake Michigan. The vertical scale is ten times the horizontal scale. (*b*) In the Gulf Islands between mainland British Columbia and Vancouver Island. The vertical scale is five times the horizontal scale.

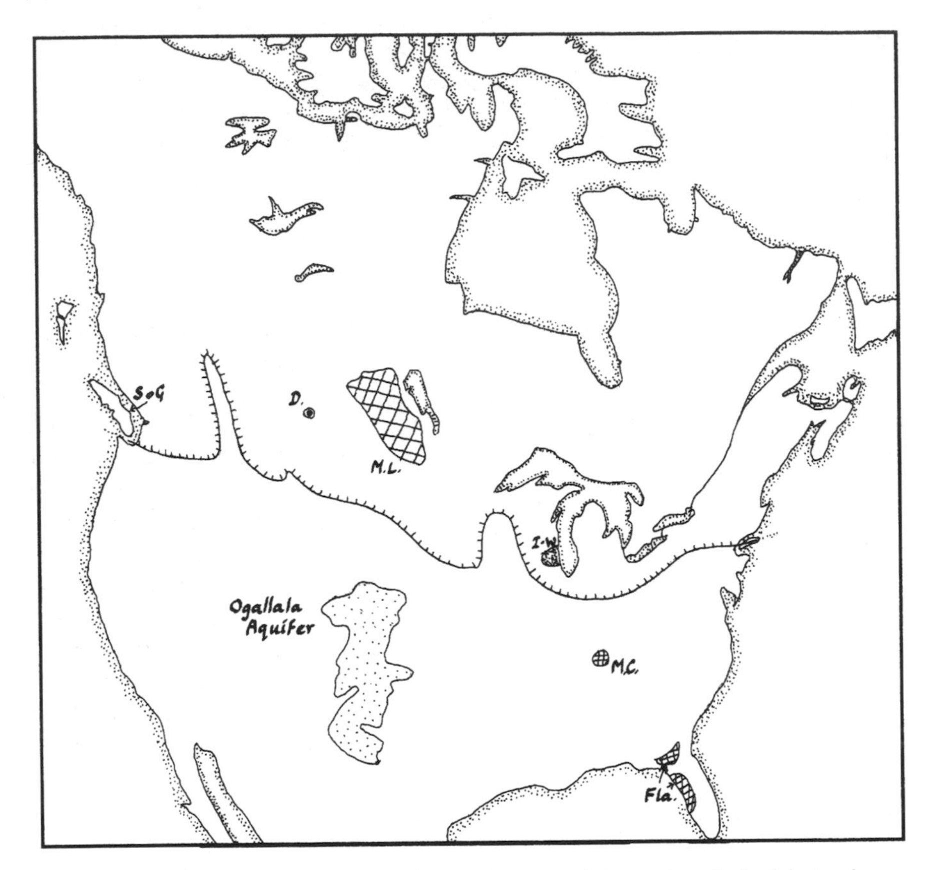

FIGURE 3.2. Map showing sites mentioned in section 3.1, and the southern limit of the ice sheets of the last ice age ( ). *S o G*, Strait of Georgia; *D*, Dalmeny aquifer; *M.L.*, Manitoba limestone; *I-W*, Illinois-Wisconsin sandstone aquifer; *M.C.*, Mammoth Caves aquifer (limestone); *Fla*, outcrops of the Florida aquifer (limestone).

is a mixture of everything from boulders to clay that was plastered down by the heavy ice. And some of it consists of clay sediments at the bottom of glacial lakes.

Whether at the bottom of a now vanished lake or forming the bulk of the till, the clay is almost impermeable; its chief ingredient is ultrafine *rock flour*. Rock flour was produced in enormous quantities by the grinding action of boulders frozen firmly to the bottom of the ice sheets, as they inched forward over bedrock. The rock flour picked up by rivers was carried into lakes; only when the flowing water finally became still could the "flour" settle to the bottom. The coarser material, boulders, cobbles, gravel, and successively finer grades of sand, settled out one after another as the rivers slowed; when in spate in early summer, the larger rivers could carry coarse material for some distance into a lake, or spread it all across a wide floodplain.

In many of the places once covered by the ice sheets (their southern limit is shown in figure 3.2), two or more layers of till now overlie the bedrock. This is because the margin of the ice sheet alternately advanced and "retreated" (melted back) over short distances, causing a given spot of ground to be alternately covered and uncovered by ice while the ice margin was in its neighborhood, sometimes several times. During each temporary advance of a few kilometers, which might last for decades or centuries, the ice would deposit a new layer of till; and during each temporary retreat, the till left uncovered—an uneven, lifeless ground surface—would acquire its own rivers and lakes, which would later be buried beneath the till laid down by the next advance of the ice. At many places, too, till layers are stacked up that date from different ice ages, separated in time by several millennia.

The result of all this ice activity was ground like that shown, in cross-section, in figure 3.3 in which three confining layers of till, dating from three separate ice advances, separate sand-and-gravel aquifers.[4] The buried aquifers are recharged in spring, while the air is still too cool and damp for the snow's meltwater to evaporate rapidly. The water collects as puddles and pools in every depression on the ground, from which it slowly soaks down, through the slightly leaky confining layers, to the aquifers below.

Big glacial-drift aquifers are much less common than small ones. One big one, nearly 1,000 square kilometers in area, is the Dalmeny aquifer,[5] just northwest of Saskatoon, Saskatchewan (figure 3.2). It is up to 30 meters thick in places, and is sandwiched between two confining layers of clayey till, a "young" layer on top and an "old" layer beneath. The young layer was laid down in the last ice age, which ended, at Saskatoon, about 12,000 years ago; possibly a fraction of the water trapped in it—it is a leaky confining layer—is meltwater from true ice-age ice. The old layer, below the sandy aquifer, dates from an earlier ice age, which ended over 100,000 years ago.[6]

On a map some of the glacial-drift aquifers have the branching pattern of river valleys; this confirms that they occupy what were once river valleys that have since become buried. The majority of the aquifers are individually small, however, just a few hectares in area and a few meters thick, capable of supplying half a dozen families perhaps, or a small farm.

Many aquifers are made of limestone. Examples of those that are impor-

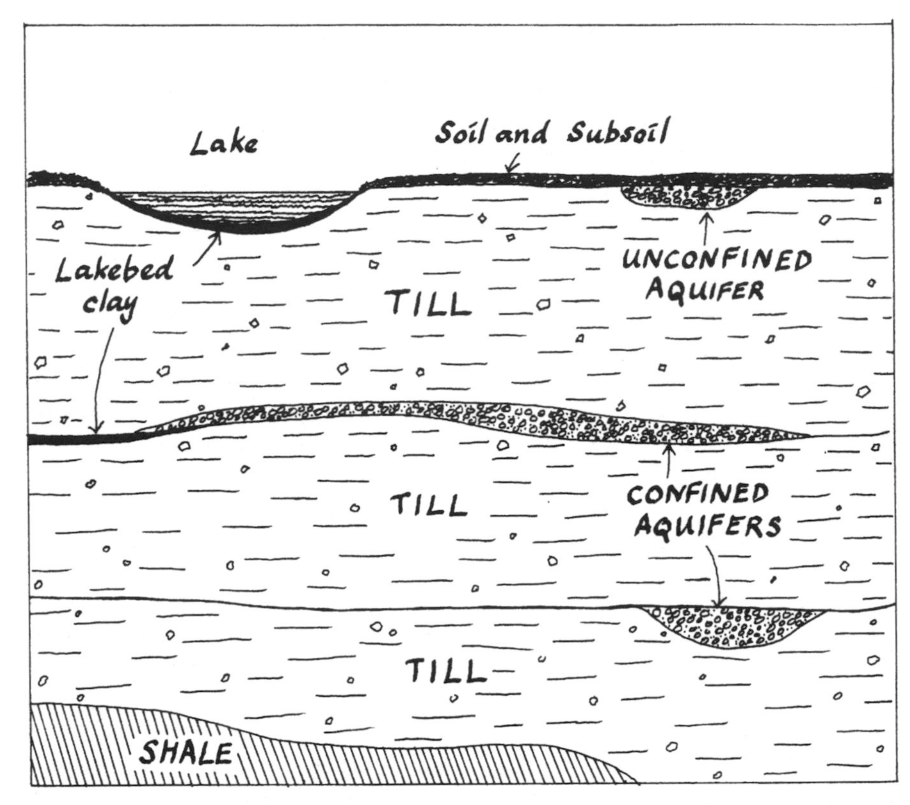

FIGURE 3.3. A typical section through the ground anywhere in prairie country that was glaciated during the recent ice ages (see figure 3.2). Three layers of till blanket the shale bedrock. A small unconfined aquifer of gravel and sand lies below the soil on the uppermost till sheet. At greater depths are two confined aquifers. The upper consists of material deposited in an ancient river valley that flowed from right to left into an ancient lake; only the clay bed of the lake remains. The lower confined aquifer is in a similar river valley, cut across by the plane of the diagram.

tant suppliers of water to their regions are big limestone aquifers in Kentucky, Florida, and Manitoba; the Florida aquifer has been called "one of the most prolific in the world,"[7] and the Manitoba aquifer is "one of the largest regional freshwater aquifers in Canada."[8]

Limestone is formed of carbonate minerals, especially *calcite* (calcium carbonate) and *dolomite* (calcium magnesium carbonate). What makes limestone so important from the hydrological point of view is that weak acids dissolve it; calcite is readily soluble, dolomite less so. Groundwater is often mildly acidic, and as it percolates through limestone, along the bedding planes and through joints and fissures, it gradually dissolves the walls of these channels, widening and enlarging them. The final result can (sometimes) be an impressive system of caves and tunnels, with fast-flowing underground rivers (for more on the rivers and streams, see section 6.7). Cave systems don't form in all limestones; in limestone that consists mainly of dolomite or that contains an admixture of sandstone or shale, the water-bearing channels in the rock remain small and widespread.

Where an underground cave system is well developed, the terrain at the surface is known as *karst.* About one-fifth of the world's land surface is karst, and in the southeastern United States the proportion is twice as great.[9] The limestone underlying karst country is known as *karstic limestone.* It forms unique aquifers. Whereas most aquifers hold water in myriads of tiny crevices through which it flows sluggishly, a karstic limestone aquifer holds water in the form of big underground lakes and rivers, with the rivers flowing as fast as surface rivers. A cave entirely filled with water because its ceiling is below the water table is known as a *phreatic cave* (from the Greek *phreatos,* a well). A cave containing a pool or river with air above it is a *water-table cave;* the water surface in the cave is, in fact, the water table. These two kinds of caves appear in figure 3.4, which shows a section through typical karst country like that of the Mammoth Cave region of Kentucky.[10]

The figure also shows a *perched aquifer,* that is, a little aquifer above the regional water table (in spite of appearances, it is not in three parts; it is merely pierced by two holes: see below). A perched aquifer forms when water is held back locally by an "island" of confining material that stops it from continuing down, through unsaturated rock, to the regional water table below. Surface water not held in the perched aquifer drains down to the deeper caves via *swallow holes,* holes in the ground giving access to the depths. They are often at the bottom of conical pits, where pools would collect if the ground were not porous.

No account of useful aquifers would be complete without mention of the most famous of them all—the Ogallala aquifer. It covers more than half a million square kilometers of the High Plains region, from Nebraska to Texas (see figure 3.2), and is a water-table aquifer of unconsolidated sand and gravel. The Ogallala is deservedly famous, both for its enormous capacity and for the way its waters are being used up faster than they are being renewed. This amounts to saying that the water is being treated as a nonrenew-

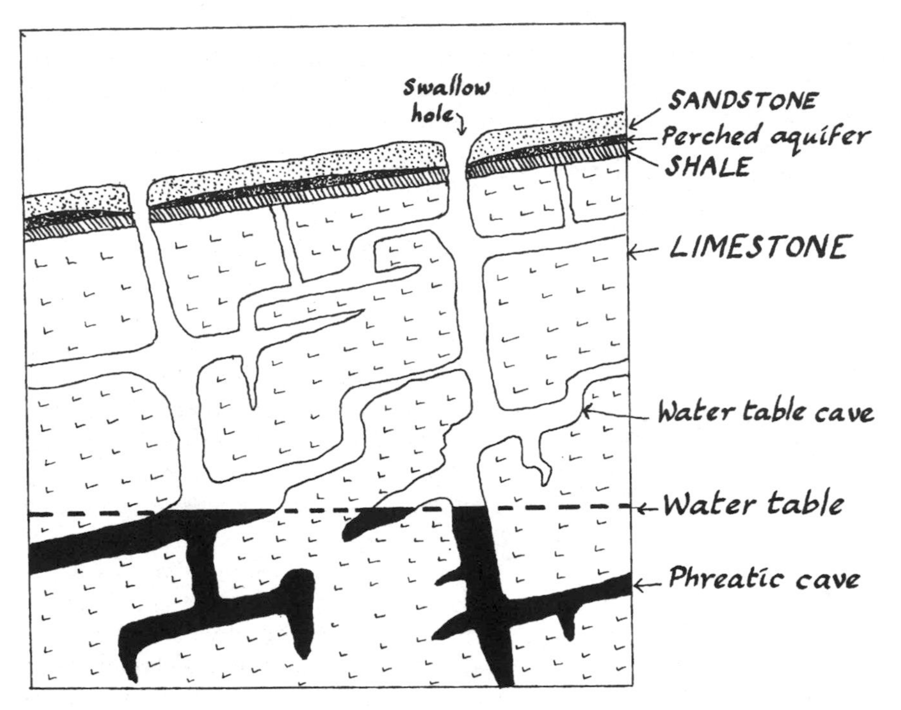

FIGURE 3.4. Aquifers in karst country. The main aquifer is a water-table aquifer in limestone. The groundwater in this aquifer is shown black. Above the limestone, and separated from it by a thin confining layer of shale, is a layer of sandstone containing a perched aquifer that slowly discharges into the swallow holes, recharging the limestone aquifer below.

able resource such as coal or oil—in a word, it is being "mined." The Ogallala is believed to contain about 4 trillion tons of water, about 20 percent more water than Lake Huron contains. In the 1970s and 1980s this water was being extracted at about 28 billion tons per year, a rate that, in theory, could be kept up for about 140 years before the aquifer runs dry. This is a gross overestimate in practical terms, because pumping the water requires more and more energy as the water table sinks (it has already fallen more than 30 meters), and would become prohibitively costly long before all the water was used up; 100 years seems a more reasonable future unless demands are reduced.[11] It has been said of the Ogallala that it "may be the greatest underground freshwater reservoir in the world."[12] It is anybody's guess how long this will remain true.

3.2 *Finding Groundwater*

Although it is true to say that an abundance of water lies below the ground in most places, it does not follow that useful aquifers are easy to find. Shallow ones are sometimes obvious—especially if water emerges at the surface through numerous seeps and springs. Deep ones are concealed, and discovering their whereabouts takes exploration. Modern hydrologists who investigate and map the subterranean "lay of the land" are exploring, just as Alexander Mackenzie or Lewis and Clark were exploring when they discovered the lay of the land in the ordinary sense, a century and more ago. Only the equipment and methods differ.

The obvious method of investigating groundwater is laborious and expensive. It is to drill test wells or boreholes and see what you find. More ingenious are the several indirect methods hydrologists use. The are easy to understand in principle, even though (not surprisingly) the details are exceedingly technical.

One method uses the fact that water-saturated rock conducts electricity more readily than dry rock does.[13]

Another method uses seismic waves, generated by a stick or two of dynamite buried in a shallow hole or even, if the bedrock is close to the surface, by the blow of a sledgehammer on a steel plate lying on the ground.[14] Seismic waves are deflected when they reach the water table and travel faster below it than above it. A seismograph, like those used for measuring earthquakes, records the travel time of the waves through saturated and unsaturated ground.

Measurements of local changes in the earth's gravity field, or in its magnetic field, can also be made to yield clues on the whereabouts of groundwater. Though uninterpretable by themselves, the clues are useful in combination with information from other sources. The "pull" of gravity (more formally, the strength of the earth's gravity field) varies minutely from place to place at ground level, depending on the density of the underlying rock. For example, a gravity study might reveal that an apparently thin sheet of loose, porous (and lightweight) surface sediments is not thin everywhere; in places it might overlie, and conceal, deeper deposits of the same material filling an old valley in heavy bedrock; and these deeper deposits *might* amount to a useful aquifer. Magnetic anomalies provide information that *may* be useful in much the same way. Sometimes aquifer-forming rocks are more magnetic than their surroundings; basalt is the most important of these rocks. Sometimes the reverse is true: nonmagnetic sedimentary rock forming an aquifer may overlie impermeable bedrock that *is* magnetic.

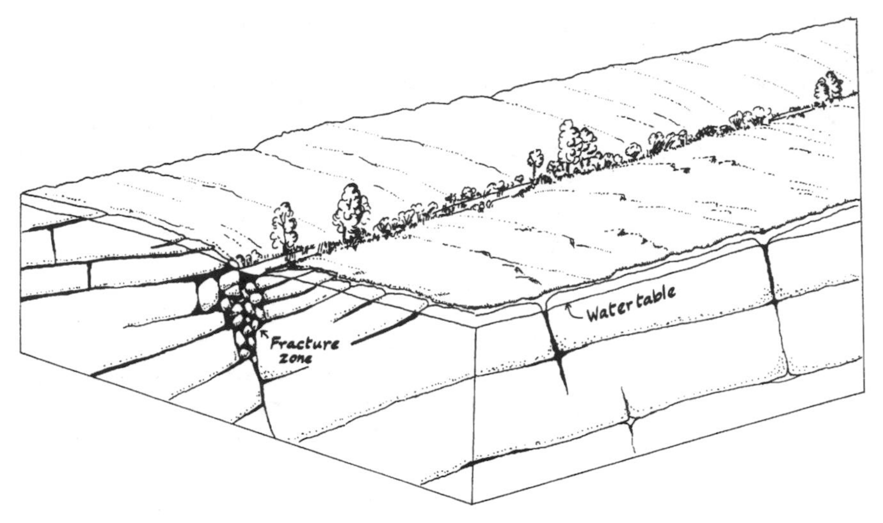

FIGURE 3.5. A fracture trace. Groundwater shown black.

These expensive, technical exploration methods are no use to the rural house-builder looking for a likely spot to drill a well. It is best to inspect the ground, using logic and whatever knowledge is available. The knowledge is often anecdotal and informal.

Inferring what lies below the ground is undoubtedly difficult. In some circumstances a useful above-ground sign of underground water is a *fracture trace.* As described in section 2.2, the water in bedrock aquifers is held in cracks (fractures) in the rock. Water is most abundant where the fractures are concentrated into corridors or zones, as they tend to be. The trick is to discover these fracture zones. The surface above a fracture zone usually subsides somewhat, as rock fragments and grains of various sizes dribble down into the cracks. The result is visible at the surface as a fracture trace, a shallow depression too straight to have been formed by a stream. A straight, dry, shallow valley may be apparent; or a straight corridor of noticeably lush vegetation; or an "unnaturally" straight stream segment, which appears to have been taken over by the stream rather than created by it (figure 3.5).

In any case, the salient feature of a fracture trace is its straightness, among a jumble of other landforms. Old, buried roadbeds, railroad tracks, and canals also make conspicuously straight lines across the landscape, as do straightened water courses and power-line clearings; the searcher for underground water has to guard against being deceived by these artificial landscape features.

Often the contrast between a fracture trace and the adjacent land is too subtle to be perceived by an observer on the ground; but the trace may show up on a pair of air photos viewed with a stereoscope, so that all the slight differences in ground texture are emphasized. Air photos also allow you to see a longer segment of a trace, making it more easily recognizable for what it is. The signs are less and less obvious the thicker the layer of soil and subsoil between the fractured bedrock and the surface. All the same, a fracture zone buried as deep as 100 meters can sometimes cause a detectable fracture trace at the surface.

By mapping, on air photos, all the fracture traces detectable over quite a large area, hydrologists can sometimes map an ancient, long-buried river and all its tributaries, which have been buried for millennia under younger sediments and soils.

3.3 Extracting Groundwater: Wells and Boreholes

Big aquifers contain huge amounts of water. There are—or were—4 trillion tons of water in the Ogallala aquifer, as mentioned in section 3.1. The water in an aquifer is never (or hardly ever) stagnant; an aquifer is forever gaining and losing water naturally (section 2.5). Over the long haul—decades or centuries—and provided no long-term climatic changes are in progress, the gains and losses remain roughly in equilibrium, unless they are interfered with.

When water is pumped from an aquifer to meet human demands, things change. One of two things happens: a new equilibrium becomes established, or the aquifer dries up: more below on the new equilibrium.

Before going into details, it's worth looking at a few numbers. Aquifers vary enormously, both in size and in the demands put on them. Furthermore, the rate at which water is withdrawn isn't necessarily proportional to the population density in the area concerned. In some regions, groundwater forms only part of the regional water supply, with the rest coming from surface water (lakes and reservoirs). And requirements vary: orchards in semidesert obviously need more irrigation water than orchards in a wet climate.

Some examples:[15] The famous Ogallala aquifer covers nearly half a million square kilometers, and the withdrawal rate is an unsustainable 50 million liters per minute, at least. In places, as much as 4,000 liters per minute can be pumped from a single big well. The Florida aquifer system covers about 200,000 square kilometers—it extends a long way into neighboring states—and the withdrawal rate (also unsustainable) is around 6.6 million liters per minute.[16] As a contrast, consider the domestic wells that supply many rural families. A small gravel aquifer in glacial drift (see section 3.1), occupying a mere 10 hectares, may yield enough water to supply several families, and do it sustainably; 500 liters per minute is regarded as adequate for a small village. Figures are necessarily vague; it all depends on what you need, and how carefully you use what you have. It has been estimated that a five-member family in a developed country uses water at the rate of 0.7 liter per minute or less, whereas a bare subsistence rate for the same family would be less than one-fifth as much.[17]

Whatever the size of an aquifer, its water behaves in predictable fashion when the aquifer is exploited. Let's consider the effect of withdrawing water from a single pumped well. First a note on the definition of the word *well.* Here it means any shaft sunk in the ground to reach groundwater, a shaft that may be dug, drilled, or driven with a pile driver. The word is sometimes used, though not here, to mean dug wells only; then the other kinds are called *boreholes.*

When water is pumped from an unconfined aquifer, the water table around the well sinks: a cone-shaped hollow, called a *cone of depression,* forms, centered on the point where the well penetrates the water table. When water is pumped from a confined aquifer (one below a confining layer), a cone of depression forms in the aquifer's piezometric surface (section 2.4). Figure 3.6 shows these two kinds of cones of depression.

The contrast between them should be emphasized. The cone in an unconfined aquifer is a cone-shaped water surface, a real, tangible surface, not an abstraction. The cone of depression in the confined aquifer is an abstraction, however. What has changed as a result of pumping is the level to which water will rise in test piezometers driven into the confined aquifer at different distances from the well. Pumping out water has reduced the head (equivalently, the pressure) of the water in the aquifer; the greatest reduction happens right at the well itself, and the effect becomes less and less marked at greater and greater distances: hence the cone shape. A confined aquifer is more "reluctant" to part with its water than is an unconfined one; it is recharged more slowly, either by leaks through the confining layers above and below it, or by horizontal flow from a distance. Some of the water it yields was held in *elastic storage;* that is, the water in the aquifer occupied space it created by

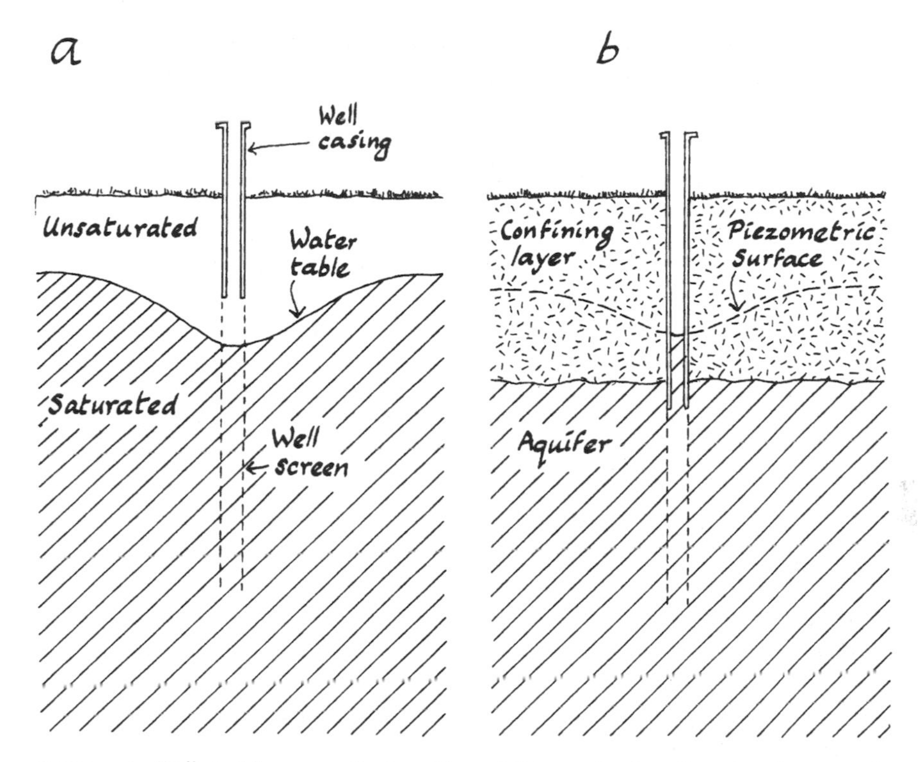

FIGURE 3.6. Wells in *(a)* an unconfined aquifer and *(b)* a confined aquifer. Note that in *(a)* the cone of depression is in the water-table surface, whereas in *(b)* the cone is in the piezometric surface, which lies within the confining layer above the aquifer.

its own pressure, which forced open the water-holding pores and cracks in the rock. When pumping begins, the pressure is relaxed, the pores and cracks contract, and the water they contained is expelled.

Note that in both the wells in figure 3.6, the upper part of the pipe is enclosed in an impermeable *well casing,* put there to ensure that loose surface material such as soil, litter, and grit won't fall into the water-filled part of the well. The lower part is walled with permeable screening, to allow as much water as possible to flow into the pipe from the saturated aquifer. Sometimes it isn't walled at all.

Figure 3.7 shows in more detail what happens when a small, household well (here in an unconfined aquifer) is pumped.[18] Assume that the pump runs

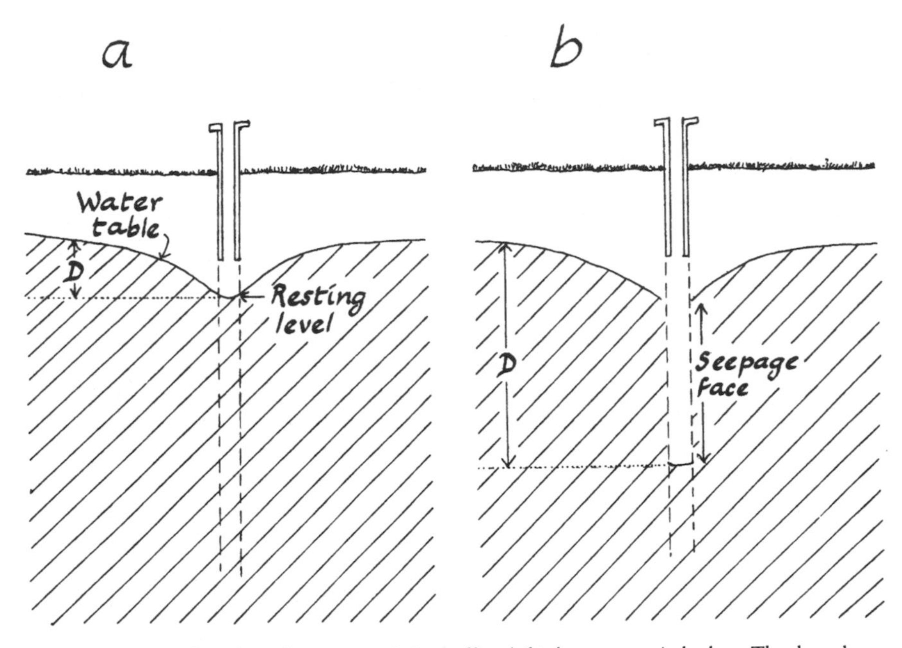

FIGURE 3.7. A well with *(a)* the pump switched off and *(b)* the pump switched on. The drawdown *(D)* is small in *(a)* when the water in the well is at its resting level. The drawdown is larger in *(b)* because the water level has sunk deep in the well, exposing an expanse of seepage face.

intermittently: it switches on automatically when the water level in a holding tank falls to a preset "almost empty" level, and switches off when the level rises to a preset "almost full" level. The pump is never off for more than a few hours, so a cone of depression is always present; it would disappear only if the pump were turned off permanently, and disappearance would take days or weeks. The water level in the well goes down and up as the pump switches on and off; that is, the *drawdown* varies (the drawdown is the difference in level between the water in the well and the water table beyond the range of the cone). While the pump is off, the water in the well is at its "resting" level at the bottom of the cone of depression. While the pump is running, the water level sinks far below the tip of the cone, exposing a stretch of the well's permeable wall, known as the *seepage face,* through which water dribbles in from the aquifer. As soon as the pump motor goes off again, this seepage refills the well up to the resting level again.

If water is extracted from an aquifer at a sustainable rate, a new equilibrium eventually becomes established. The flow net will be radically different from what it was before, because some of the aquifer's discharge will leave via artificial wells. But *if* equilibrium is attained, discharge and recharge will balance each other again.

The course of events at a single well, from the time pumping first starts until the time a new equilibrium is reached, goes like this (we disregard the rapidly alternating, small-scale changes that happen as the pump motor switches on and off). The well's cone of depression steadily grows deeper and wider, until its outer limit reaches a discharge or a recharge area. If it reaches a discharge area, the rate of natural discharge will slow down. If it reaches a recharge area, recharge will increase. The cone itself acts as a funnel; rainfall that previously flowed away over the surface can now sink into the ground.

As long as discharge exceeds recharge, the cone keeps on growing. If all goes well, a time will come when the recharge rate is great enough to balance the total discharge rate—natural plus artificial. Once this happens, and it can take years, the cone of depression stops growing; equilibrium has been attained. In a densely populated area with fields of wells, the same things happen on a larger scale. The only difference is that in place of a single cone of depression, there are numerous overlapping cones, and the geometry becomes correspondingly complicated.

All is not well, of course, if equilibrium is *not* reached or, equivalently, if water is extracted at an unsustainable rate, faster than it can come in; or to use yet another phrase, if the water is "mined." This is what is happening in the Ogallala aquifer. It is also happening in the sandstone aquifer lying across the Illinois-Wisconsin border, the water source for millions of people, among them the populations of Chicago and Milwaukee (see section 3.1 and figure 3.1).

Figure 3.8 shows how 90 to 100 years (approximately) of pumping has affected the piezometric surface of this aquifer.[19] The figure gives a north-south profile of the earlier and later piezometric surfaces only; the ground surface and the boundaries of aquifers and confining layers have been left out to avoid overcrowding the picture (but see the east-west section in figure 3.1a). The upper line in figure 3.8 shows the piezometric surface as it was in the latter half of the nineteenth century, the lower line as it was in 1973. Two huge cones of depression have developed, centered on Chicago and Milwaukee. A high groundwater divide, not present in the past, separates the cones. The vertical scale is exaggerated; it is more than 650 times the horizontal scale. Even so, the change is remarkable and it is presumably still increasing.

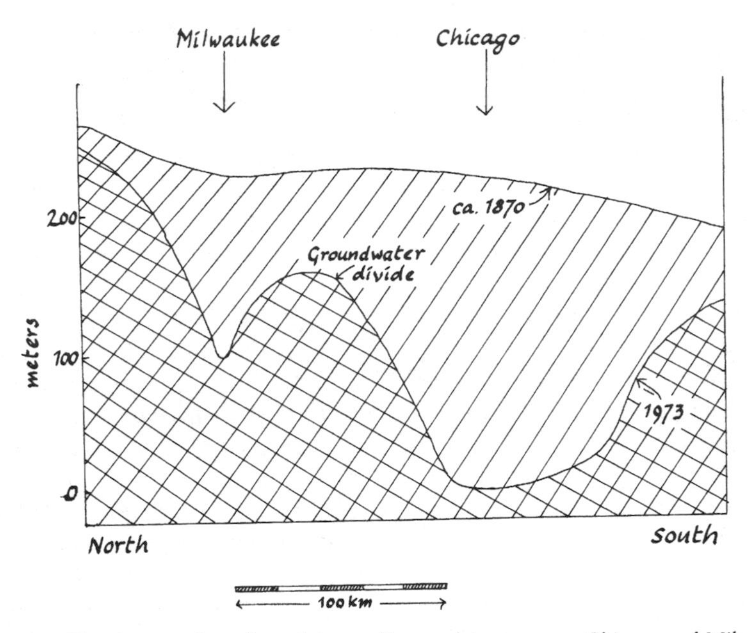

FIGURE 3.8. The piezometric surface of the aquifer supplying water to Chicago and Milwaukee, as it was in the 1870s and in 1973. This aquifer is the same one shown in figure 3.1a. Geological details are omitted here for clarity. Note the cones of depression that have developed, centered on the two cities, and the groundwater divide between them.

To quote an authority writing in 1988: "Unless groundwater withdrawals are reduced . . . water levels will continue to decline until the aquifers can no longer transmit water . . . to the well fields."[20]

Extracting groundwater from the ground does more than cause an invisible lowering of the water table (if the aquifer is unconfined) or the piezometric surface (if the aquifer is confined). The water becomes progressively more expensive to pump and more contaminated with dissolved minerals. Extraction brings about visible changes in surface waters, too. The flow of rivers and streams may be reduced; ponds, marshes, and bogs sometimes dry up; near the coast, freshwater aquifers become contaminated with sea water, which seeps in to replace the fresh water pumped out. The most spectacular result is the subsidence of the land itself; this can happen when an aquifer's cavities, filled with air instead of water, crumple and collapse because they are no longer internally supported. This has happened in many places, for example in California's San Joaquin Valley and around Phoenix, Arizona.[21]

These disasters don't always happen; the land is forgiving, up to a point. But as the population increases, no doubt that point will be passed more and more often.

3.4 How Fresh Is Fresh Water?

Groundwater is a useful source of fresh water only if it is sufficiently pure. A lot of it isn't. Water at depths greater than a few hundred meters is likely to contain such a high concentration of dissolved minerals as to be unusable. This means that shallow wells (but not too shallow; see section 3.5) usually yield better water than deep ones. It also explains why a rural family whose well once gave good, pure water finds water quality deteriorating as the population in their neighborhood increases; as nearby wells multiply, the hollow formed by their combined cones of depression expands and deepens, and the wells draw deeper, more mineralized water.

When it comes to the minerals concerned, there is a striking contrast between groundwater and soil water, because the chemistry of soil water is strongly influenced by the plants and the myriads of other living organisms in the soil; many of their by-products never percolate down into groundwater.

The principal metals in groundwater are calcium and magnesium, the basic ingredients of all kinds of limestone; and sodium and potassium, lightweight, chemically active metals; sodium is the metal that, combined with chlorine, forms common table salt, the chief ingredient of sea salt. Chlorine is one of the principal nonmetals in groundwater. The others are sulfate, a combination of sulfur and oxygen; and carbonate,[22] a combination of carbon, oxygen, and (often) hydrogen that, combined with calcium and magnesium, forms limestone. When groundwater is analyzed by a hydrologist, a list of the amounts of these chemicals in the water is what you get. They are the "standard" groundwater chemicals; sometimes, as we shall see below, nonstandard chemicals turn up as unpleasant surprises.

If you are a well owner, you get an entirely different list of chemicals when your water supply is analyzed by health officials. The standard chemicals are of little interest from the health point of view, but others, present in relatively tiny amounts, are. Examples: fluoride, which in small quantities protects children's teeth, but in large amounts discolors them; barium, which can damage the heart, blood vessels, and nerves; boron, which attacks a whole range of organs, notably the kidneys; nitrate (nitrogen combined with oxygen), which affects a baby's ability to absorb oxygen; and, of course (though these are hardly minerals!) fecal coliform bacteria. But the list would almost certainly not mention many of the exceedingly toxic industrial chemicals that occasionally pollute well water, because they cannot be detected by routine tests. More on these chemicals in section 3.5.

Still another list of chemicals mention those that interest farmers and gardeners. A gardener buying fertilizer knows that the indispensable elements for a garden are nitrogen (in nitrate or ammonia), phosphorus, and potassium, or N, P, and K, to give them their chemical symbols as listed on the labels on bags of fertilizer. These are soil chemicals, however. Except for potassium they rarely occur in significant amounts in groundwater: phosphorus hardly ever, and nitrogen only occasionally.

Much groundwater contains natural contaminants and is classed as "poor-quality groundwater." Some examples: In many places in Atlantic Canada, well water is contaminated with arsenic in concentrations dangerous to human health; the arsenic comes from local rocks. Another contaminant from the rocks, fairly common in the neighborhood of coal mines or oil wells, is hydrogen sulfide, the gas that gives rotten eggs their characteristic smell; the sulfur comes from coal or oil. Hydrogen sulfide is merely an unattractive water ingredient rather than a threat to health; although it is deadly poisonous at high concentrations, oddly enough its smell becomes undetectable at concentrations high enough to be dangerous.

In much of Canada north of the Great Lakes, and along the British Columbia coast, the groundwater at depth is often contaminated with common salt that is probably of very ancient origin; it is "relict" salt, left behind by ancient seas that once covered the land, hundreds of million of years ago in the case of central Canada. Groundwater can also become contaminated with present-day sea water as a result of human activities. Another ancient contaminant is found in many wells in southern Alberta; it is nitrate, believed to have been formed in the soil (from ammonia) as much as 3,000 years ago, when the climate was warmer and drier than it is now.[23]

A comparative measure of the purity of a groundwater sample is its *TDS*, which stands for total dissolved solids; the amount is measured in milligrams per liter (mg/l), which is the same as parts per million (ppm). Pure water contains less than 100 mg/l of dissolved solids; this is equivalent to 17 grams (about 2 teaspoonfuls) in a full bathtub (about 170 liters). Acceptable drinking water must contain less than 500 mg/l; livestock are expected (by law) to put up with five times as much.[24]

The chemicals in groundwater are always moving. As water flows through an aquifer from a recharge to a discharge area, it picks up chemicals on the way, with the result that the TDS in discharge water is often several times as great as it was in the recharge zone. The slower the water flows, the more time there is for chemicals in the rocks to dissolve, and the higher the TDS becomes.

A confined aquifer also acquires chemicals from the confining layer above it, if the layer is leaky. The leaking water picks up chemicals on its way down and carries them into the aquifer. For example, the confining layer of glacial till over the Dalmeny aquifer near Saskatoon contains much gypsum, the mineral used to make plaster of Paris. Spring meltwater recharges the aquifer. The meltwater collects in pools, wherever there's a surface depression, so the descending water travels down (at a rate of a few millimeters per month) in a separate column below each pool, dissolving gypsum on the way. As a result a number of distinct, gypsum-free "chimneys" penetrate the confining layer.[25]

As a general rule the deeper, and older, the groundwater, the more chemicals it contains—in other words, the greater its TDS. This is not true, again as a general rule, of oxygen. Usually, much of the oxygen dissolved in rainwater disappears before the water sinks far into the ground; some is used by soil organisms as the water percolates through the soil; what remains combines chemically with minerals before it has time to reach any great depth; as a result, deep groundwater is usually devoid of oxygen. But not invariably. Sometimes the material the groundwater percolates through is incapable of absorbing oxygen because it already contains all it can hold; when this happens, oxygen remains in the groundwater. In parts of Nevada and Arizona, water 1,000 meters below the ground, 80 kilometers from the point where it entered the ground, and more than 10,000 years old has been found to be well oxygenated.[26]

A form of contamination that happens only near the seacoast is saline water contamination. Although the contaminant—the saline groundwater below the seafloor—is natural, it finds its way into freshwater aquifers as a result of human activities.

Aquifers near the coast slope down to seaward. In an aquifer that lies across (and buried beneath) the coastline, the groundwater is fresh at the landward end and salty at the seaward end; the freshwater-saltwater interface is quite abrupt and is usually close to shore. Its position is governed by the rate at which the aquifer's fresh water is being discharged seaward; the seaward flow of fresh water prevents the landward flow of salt water. As long as the climate and sea level don't change, the interface remains in a natural equilibrium position, except for minor cyclical changes going on all the time, caused by tides and seasons. But if water is pumped from the freshwater side, the interface shifts, and the result is *saline-water encroachment* into the aquifer. In minor encroachments all that happens is a slowing down of the outward flow of discharging groundwater and a landward shift of the interface to a new equilibrium position. More serious encroachments occur when the natural direction of flow, on the seaward side of the wells, is reversed; this happens if the wells create a cone of depression so large it captures all the freshwater discharge and sucks in salty groundwater, from under the sea, as well. The saline water *upcones* below the wells (figure 3.9).

Although the freshwater-saltwater interface in unconfined aquifers is seldom far from shore, the groundwater in confined aquifers below the seafloor may sometimes be fresh for a long way out to sea. For example, off New England the water in confined aquifers 200 meters below the seafloor is still fresh 60 kilometers out from the coastline.

3.5 Polluted Groundwater

The words *contaminate* and *pollute,* as applied to water, are usually taken to mean the same thing.[27] Their use as synonyms, however, prevents what could be a useful distinction: if, as in this book, we use *contaminated* to mean naturally tainted, and *polluted* to mean tainted because of human activities, a frequent source of misunderstandings disappears.

Pollution, in its many forms, is a serious threat to the world's freshwater supplies. The pollutants are chiefly sewage in the underdeveloped world and industrial effluents in the overdeveloped world, but this is not to say that sewage is never a problem in the latter: far from it.

Pollution is commonly divided into two kinds:[28] that coming from single, identifiable, point sources; and that coming from distributed sources spread over extensive tracts of land. Common point sources are leaky gasoline tanks at filling stations, leaky municipal sewage lagoons, municipal landfills, feedlot effluent, mine spoils and tailings, factory effluents, stores of road salt used on icy roads in winter, accidentally ruptured road and rail tankers, and septic tanks and tile fields. The commonest distributed sources are the pesticides and fertilizers spread by farmers, logging companies, golf-course managers, and suburban gardeners; runoff from roads; runoff from power-line routes kept clear with weed killers; and leaky municipal sewer lines.

The aquifers most at risk are unconfined (water-table) aquifers. Surface water drains down to them quickly and directly. When a polluting liquid

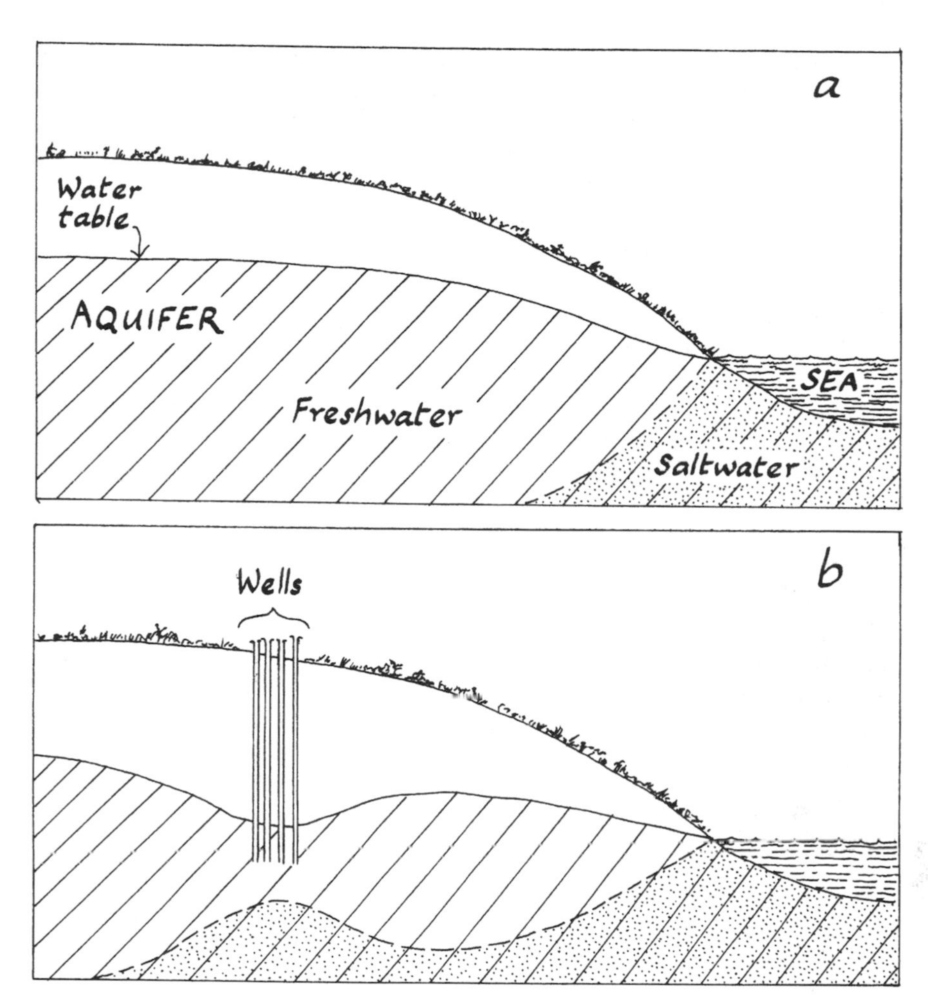

FIGURE 3.9. An unconfined aquifer at the seacoast *(a)* in its natural state and *(b)* with a field of wells pumping water from it. Note the upconing of the saline groundwater below the wells.

reaches the water table, one of three things happens: if the liquid is water with the pollutants dissolved in it, it mixes with the groundwater and becomes part of it; if the liquid is lighter than water, for example, gasoline, it floats as a pool on top of the water table; if it is a heavy liquid, it sinks through the aquifer to form a pool resting on the confining layer at the bottom.

Septic tanks, landfills, and agricultural fertilizers are the most widespread sources of what, not to mince words, is foul water. The deeper the water table, the better the chance that some of the pollutants will be filtered out, and that biodegradables will biodegrade, before they reach the groundwater. Among the pollutants from septic tanks and sewage lagoons are disease-causing microbes, and chemicals from dishwasher and laundry detergent. Making the chemicals foamless, odorless, and tasteless isn't necessarily beneficial. Detergents that cause water to foam, or have a recognizable taste or smell, can alert a well owner to septic-tank pollution that might otherwise go undetected.

Municipal landfills (figure 3.10a) contain a mixture of pollutants. Rainwater dribbling down through the landfill washes them out, and the resultant liquid is known as *leachate.* It is a soup of decaying garbage, waste building materials, sludge from wastewater treatment plants, and numerous other ingredients, including decomposed dead rats.

Leachates also come from the spoil piles and mill tailings at a mine site; if untreated they are often strongly acid and are likely to contain toxic metals.

Agricultural fertilizers (this includes manure, as well as chemical fertilizer) pollute groundwater with nitrate; sometimes this is removed by bacteria living in the aquifer, in a process known as *denitrification.*

So much for "watery" pollutants. Liquids that don't mix with water are also common pollutants; as remarked above, some are lighter than water, some heavier.

The commonest of the lightweight polluting liquids is gasoline. Sometimes it seeps into the ground from a leaky underground tank at a filling station (figure 3.10b); sometimes it gushes from a tanker following a road or railroad accident. It soaks down through the soil and subsoil until it reaches saturated ground. There it spreads to form a pool floating on the water table. If that were all, it would be harmless. But it is not all. Gasoline contains a number of toxic compounds that dissolve out of the gasoline into the groundwater below. Many of these compounds, for instance benzene, are dangerously poisonous at concentrations too low for their presence to be detected by taste or smell.

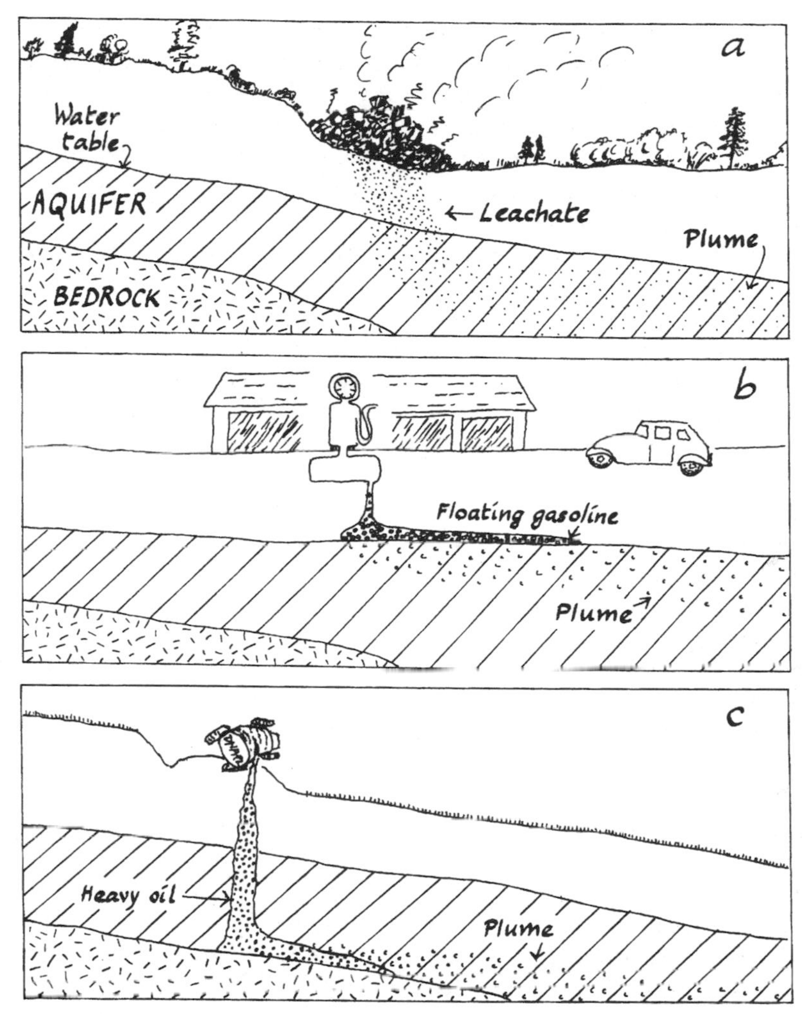

FIGURE 3.10. Three kinds of pollution entering an unconfined aquifer. *(a)* Watery leachate from a landfill, mixing with the groundwater. *(b)* Leaking gasoline. Pollutants dissolve into the groundwater from the floating pool of gasoline. *(c)* A spill of a heavy, oily pollutant (DNAPL). Polluting chemicals dissolve into the groundwater from the pool of DNAPL resting on the bottom of the aquifer.

The pollutants that dissolve into groundwater from a floating pool are carried a long way, sometimes many kilometers. The *plume* of pollution from a continuously flowing pollution source, such as a gas tank with an unsuspected leak, lengthens out like the smoke plume from a factory chimney. A single accidental spill produces a single *slug* of pollution, which moves with the groundwater as a unit, gradually becoming larger; eventually, it becomes so large, and the pollutants so dilute, that they are no longer harmful.

The heavy polluting liquids behave in much the same way, except that they sink through an aquifer and pool at the bottom of it instead of floating at the top (figure 3.10c). Most are oily industrial solvents that are less viscous than water and sink quickly. They are known collectively as *dense nonaqueous phase liquids* or DNAPLs for short. Two familiar examples of the many different kinds are carbon tetrachloride and trichloroethylene. Very tiny spills of such chemicals can do widespread, long-lasting damage. For example, the trichloroethylene in a standard 200-liter drum would need to be diluted with 60 billion liters of water to make it harmless.[29] It would take almost 230 years for this amount of water to flow from a well supplying 500 liters per minute, a rate capable of supplying a small village.

Most North Americans have no control, as individuals, over their water supply, but they can reasonably expect its quality to be carefully monitored. The rural minority who do exercise control over their water, that is, the owners of private wells, have to do their own monitoring. The pollution sources most likely to affect them are septic tanks, fertilizers, landfills, and (on farms) private fuel tanks. The aquifers most likely to be polluted are water-table aquifers; polluted water flows down into them directly, and if the water table is shallow, there is not much time for the pollutants to be absorbed or degraded before they enter the groundwater. Sometimes a well owner can drill down to a lower, confined aquifer, which is much less likely to be polluted. When this is done, however, it is crucial to ensure that the well casing extends to below the polluted aquifer, as water pumped from the well will consist of a mixture from all the aquifers in contact with the well's seepage face.

Confined aquifers are not proof against pollution, even though they are less at risk. Sometimes cracks and fissures in a confining layer lets polluted water flow down rapidly into an aquifer below. The confining layer may be fissured bedrock, or a till layer of stiff clay, which tends to crack as it dries. Pollution can also penetrate, slowly, through unfissured, very slightly permeable, confining layers. Indeed, in a great many places it has been doing so for decades.

Slowly moving pollution is the greatest threat to our water supplies at the present time. Nowadays, care is taken (or ought to be taken) to stop pollutants at the surface, but much pollution is already below ground level, beyond reach. Some of it, as you read this, is gradually seeping through leaky confining layers. And some of it is still in the process of being dissolved out of old pools of DNAPL, and will need to be diluted by enormous quantities of pure water for it to become safe.

The most dangerous of the hidden pollutants are probably those dating from the post–World War II period that are still underground, en route to as yet unpolluted aquifers, especially if they come from unreported spills at long-forgotten locations. As one authority on the topic has put it, pollution of this kind "is a problem of increasing magnitude. Cleanup of much of this contamination is not feasible with existing technology."[30]

4

Water below the Ground: Vadose Water

4.1 Underground Water above the Groundwater

When rain falls on dry land, or when snow melts on it, the newly arrived water doesn't become part of the groundwater immediately. Before it can do so, it must penetrate the unsaturated ground overlying the water table, known as the unsaturated zone. This may take considerable time, and not all of the incoming water necessarily gets as far as the water table; after a temporary sojourn in the unsaturated zone, some of it usually returns to the atmosphere again. The way water behaves in the unsaturated zone is the topic of this chapter.

Its "behavior" amounts to movement, up and down and horizontally; it is seldom at rest. We shall follow its ceaseless comings and goings and ask how and why it moves, how fast, through what channels, and where to.

First, a few technical terms. The unsaturated zone has two other names: sometimes it is called the *zone of aeration,* because its pores usually contain some air, and sometimes the *vadose zone* (from the Latin *vadosus,* shallow). From this point on, we shall use the word *vadose;* its brevity makes up for its being a jargon word, and it is more accurate, too; the vadose zone is not unsaturated always and everywhere, as we shall see. Nor is it a single, homogeneous zone: in most places it has three distinct layers. The topmost layer is the soil; below it is the *intermediate zone,* consisting of the subsoil together with any solid bedrock that happens to lie between the subsoil and the water table; and at the bottom, in contact with the water table, is the *capillary fringe,* a layer into which groundwater soaks upward, like water into a sponge.

The soil is the most familiar, and certainly the most complex, of these layers. The distinction between soil and subsoil is that the former contains living organisms larger than bacteria while the latter does not. The soil is the layer occupied by the roots of plants; it is also the home of myriads of mostly unnoticed soil organisms. This abundance of living material affects the soil's structure, its color, its chemistry, and the way water flows through it. Indeed, when water descends from soil into subsoil, it encounters entirely different conditions.

The thickness of the soil varies enormously: in some places it is no more than a thin layer of dust on bare rock; in others, it may consist of peat beds 4 or 5 meters thick.[1] Soil is continually subject to two opposing forces: dead organic material, accumulating ceaselessly, continually builds it up; at the same time never-ending erosion, as water flows through and over it, continually washes it away. If the soil thickness remains unchanged for long periods, as in many places it does, the reason is that these two processes balance each other.

The intermediate zone is the zone of loose mineral sediments, and sometimes bedrock, sandwiched between the soil and the capillary fringe. In places where the soil is deep or the water table shallow, there may be no intermediate zone; then all that lies between the surface and the capillary fringe is true soil.

The two upper layers of the vadose zone together form a temporary holding ground for water on its way to some other destination: some flows down to join the groundwater, some flows laterally to adjacent streams and lakes, and some is transported back into the atmosphere by the action of plants (see section 4.5).

The lowermost of the three layers of the vadose zone is the capillary fringe. It is not necessarily different in texture from the material above and below it, but it is permanently wet, saturated with water sucked from the groundwater beneath. The height to which water will rise above the water table because of this spongelike suction depends on the texture of the material holding the water. The water moves up to fill the pores and interstices of this material and the tiny channels that link them. It is constantly being pulled downward by gravity at the same time as it is being pulled upward by the suction force of surface tension; the level at which these opposing forces balance—the outcome of the tug-of-war, so to speak—determines the thickness of the capillary fringe. Water rises much higher in fine-grained material like clay than in coarse-grained material like sand; this is because the narrower the crevices holding the water, the less the weight of each individual "thread" of water, and therefore the higher it can be pulled upward by surface tension.

The result is that the thickness of the capillary fringe ranges from a millimeter or less in the coarsest sand to as much as 2 meters (or even more) in very fine clays. In any case, the *capillary water,* as it is called, has an irregular upper surface because of the irregular shapes and sizes of the pores it fills; hence the term capillary *fringe* rather than capillary *layer.* In places where the water table lies at no great depth beneath a fine-textured soil, the capillary fringe can reach right up to the surface of the ground. When this happens, groundwater evaporates directly into the atmosphere. As the water table rises and falls, the capillary fringe rises and falls with it. When the water table shifts up and down in shallow soils, the top of the soil alternates between being moist and cold while the capillary fringe is at the surface, and comparatively warm and dry while the fringe is below the surface. The alternations can be detected from the air, using remote sensing devices that are sensitive to temperature changes in distant objects.

A point that may seem puzzling at first thought is this: if the capillary fringe is saturated with water, and if the water is all of a piece with the groundwater, why is the fringe regarded as a layer of the vadose zone, which, by definition, lies entirely above the water table? Why don't hydrologists redefine the term *water table* to mean the top of the capillary fringe? These questions are answered in section 4.4.

4.2 *From Sky to Soil*

Precipitation (rain and snow) obviously has to reach the soil before it can penetrate it. Before considering how water behaves once it is in the vadose zone, we must give a moment's thought to what happens to precipitation on its way down. An appreciable fraction of it is "caught" by vegetation and evaporates without ever reaching the ground.

This is especially true in forested country. Recall the last time you sheltered from a rainstorm under a tree's spreading canopy, or noticed the pale circles of unwetted ground below the trees bordering a road when all other surfaces were drenched and darkened by heavy rain. Obviously, not all the rain that fell has reached the ground. Similarly in a snow-covered forest in winter. Below each snow-laden tree is a deep, cup-shaped hollow partly caused by the interception of snow by the tree's crown. Melting of the snow where it touches the dark, sun-warmed tree trunk makes the hollow even deeper (see figure 4.1).

In both cases the precipitation has been delayed on its downward journey; what happens next? When rainfall is heavy enough, some of the rain wetting a tree will eventually drip to the ground or dribble down the tree's trunk; and some of the snow caught on a tree's crown in winter will either slide off as snow or drip off when it melts. But in both cases a fraction of the water or snow goes back into the atmosphere without ever reaching the ground: the water is evaporated, or the snow *sublimated* (vaporized directly without

FIGURE 4.1. Interception of snow. Note the deep hollow at the foot of the tree, largely the result of interception.

first melting to liquid water). The result is that a substantial amount of the water that falls from the sky as precipitation never has a chance to soak into the ground; to use the technical term, it is *intercepted* before it ever gets there.

Interception is an important event in the hydrological cycle of places with wet climates. The amount of water involved is often appreciable.[2] For example, of the rain that falls on coniferous forests in summer, between 20 and 40 percent is usually intercepted. When rain falls gently for long periods, water evaporates from the wetted leaves even while the rain continues to fall; consequently, the amount intercepted by a tree is often much greater than the quantity of water the tree can hold at a single moment. The total amount intercepted may amount to several complete "wettings." Tall grass is nearly as important as trees in influencing the hydrological cycle; it often intercepts between 20 and 30 percent of summer rain.

Next consider the precipitation that runs the gauntlet of the interception process and actually wets the ground. Even this water will not all soak down below the surface; some will evaporate immediately. A lot of it will soak in, of course, and if the rain is heavy enough, or the snowmelt fast enough, the topmost layer of the soil will become saturated and unable to absorb more water. (In other words, part of the "unsaturated" or "aerated" zone has become saturated, albeit only temporarily: hence the reason for preferring the name *vadose zone.*) It is often obvious when this has happened: in warm weather, earthworms come to the surface to avoid the flood, and lie there drowned. If heavy rain persists, or snow continues to melt, the water will collect on the ground, first creating puddles in every surface hollow and then flowing away, as *overland flow,*[3] to join the nearest body of surface water.

4.3 *How Water Enters the Vadose Zone*

After interception, evaporation from the surface, and overland flow have taken their toll, the precipitation remaining makes its way into the soil and soaks down into it. The process is known as *infiltration.*

Water soaks easily into "good" soils, the kind that farmers and gardeners find most productive. This is the kind of soil that crumbles readily; water dribbles slowly through the pores and gaps separating the crumbs, gaps that owe their existence to the host of invertebrate organisms that spend their whole lives in the soil. These exist in huge variety.[4] The biggest are earthworms, centipedes, millipedes, and tunneling spiders. Besides these (see figure

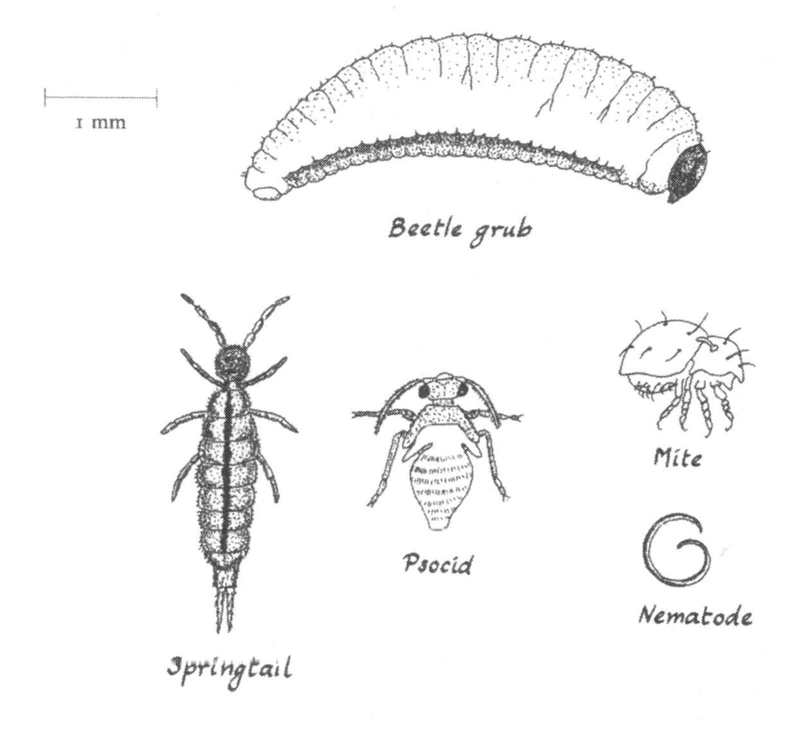

FIGURE 4.2. Some small soil animals. Note the scale mark at the top.

4.2) are the grubs of various beetles, and an array of mites and smaller insects, ranging down to minute springtails and psocids (relatives of the book louse). Most plentiful of all are nematodes, tiny, smooth worms, mostly invisible to the naked eye, that seem never to stop rapidly bending and straightening, coiling and uncoiling. They have been described as "inconceivably abundant in moist soils" and as looking like "animated bits of fine sewing thread."[5]

In addition to this multitude of quite highly developed animals, there is another multitude of microscopic one-celled organisms: protozoa such as amoebas, plus numerous algae, bacteria, and yeasts. Any drop of water seeping down through soil is jostled by hosts of tiny organisms.

Larger animals also play their part. The burrows of vertebrate animals—mice, voles, moles, gophers, rabbits, ground squirrels, and shrews—all function as water channels. Other channels are created where the roots of trees and smaller plants die and decay. Tiny channels form where fungus filaments grow below the surface; these are especially abundant in forest soils, where the filaments appear as networks of delicate white threads.

Besides flowing through all the various channels made by living things, soil water also flows through the cracks that form in sunbaked soil as it dries, and through the crevices left by melting ice crystals as winter gives way to spring. Needle ice (see section 8.4), which forms big sheaves of close-packed "needles" just under the soil surface, often leaves a honeycomb of hollows, some as big as a matchbox.

So much for the factors that open up the soil, speeding the flow of soil water. Now consider the many factors having the opposite effect. An important one is soil compaction, which can have many causes: pressure from wheels is the most obvious; pressure from animals' hooves and even the battering of heavy raindrops also contribute. The flow of water is hindered when, after any lengthy dry spell, a surface layer of windblown dust is washed down into the soil pores by the first rain; besides being clogged, the newly wetted pores are also likely to shrink because their walls absorb water and swell. Heavy rain can turn firm clods of soil to liquid mud, which flows down to block the channels below. In cold climates, ice crystals that fill the soil cavities in winter melt later than the surface snow, and prevent the snow's meltwater from soaking into the ground in spring. And in some soils, minerals and fine particles leached from the topmost layers react with each other chemically and "set" to a cementlike hardness, forming an impermeable *hardpan* below the surface.

Vegetation does much to protect the soil and make infiltration easier. It protects the soil surface from compaction by heavy rain; it slows overland flow, giving the water more time to be absorbed; it creates breaks in the soil surface through which water can enter; and a layer of leaf mold, besides contributing to the cushioning effect that prevents soil compaction, also filters out fine particles that would clog the soil channels.

Unless the water table is very shallow, water that has traveled down through the soil reaches the intermediate layer of the vadose zone. Here it continues its journey in entirely different circumstances. Subsoil is usually less permeable than soil because it is devoid of growing, breathing, moving, burrowing, and wriggling organisms. Its only channels are the interstices among mineral grains and the cracks formed by drying or freezing.

The manner in which water travels, down and sometimes up, in the soil and unconsolidated subsoil is the topic of section 4.4. Water even flows sideways, if it is diverted from its downward path by a layer of hardpan. The way water in huge quantities is captured by plants, transported upward, and lost into the air, is considered in section 4.5.

The lower part of the intermediate layer sometimes consists of unsaturated bedrock. Bedrock is often plentifully supplied with fissures and fractures and joints which, if they are above the water table, contain at least some air. Much of the time they also contain driblets and streamlets of descending *vadose water* (the water in the vadose zone).

In mountainous country, fractured bedrock often forms the surface of the ground—soil and subsoil are missing. Seemingly dry valleys mark the course of underground streams. Often the dry rocks making up the floor of such a valley are arranged in what looks like a flat, neatly constructed "pavement"; this is because, over many years, a subterranean stream has rinsed away the mud and gravel among the heavier rocks, leaving the latter to settle and become fitted to one another, side by side. And sometimes a stream is found that appears to go up and over a saddle (!) to link a pond-filled hollow on a mountainside with the slopes below (figure 4.3). What is happening is that

FIGURE 4.3. The channel leading out of this pond seems to go upward over a saddle. In fact, the stream draining the pond flows away below the surface, undermining the soil over it, which has collapsed to form a small valley.

the pond is draining through an underground stream (flowing downhill, naturally) that has undermined the soil above it; the soil has collapsed to some extent but not far, so that the valley resulting from the collapse begins with an uphill section.

In mountainous country with shallow soil, runnels between the soil and the underlying solid bedrock give water a fast, invisible route down the mountainsides. Runnels can also form under the peat of peat bogs, and drain water away down much gentler slopes; the water is invisible, but its presence is revealed by the darker green of the overlying moss.[6]

By far the biggest of the channels in the vadose zone are the caves in karst country (regions with cavernous limestone). Consider the cave system shown in figure 3.4. It shows three kinds of caves: those entirely filled with water, right up to the roof, which are known as phreatic caves; those containing rivers and pools, with air above the water, which are known as water-table caves because the water surface *is* the water table; and dry caves at a higher level. These last are unlabeled in the figure; they are technically known as *vadose caves* because they are in the vadose zone; the water that flows through them after a heavy storm at the surface is, while in the zone, vadose water. Cave explorers following an underground river are visitors in the vadose zone.

4.4 *How Water Moves in the Vadose Zone*

Imagine the rain from a heavy storm soaking into "dry" soil. (Dry soil is not totally devoid of water, as we shall see below.) The water seeps down through the numerous channels described in the preceding section, as a "wave" of moisture; to begin with, this wave has a sharply distinct lower surface—the *wetting front.* Two forces are pulling the water down: gravity and the suction of the dry soil which, in this case, pulls downward as well. If the soil is sodden with water, the force of gravity is far greater than that of suction. In comparatively dry soil the opposite is true. As the water from a short, sharp rainshower sinks into the soil, it spreads through an ever-expanding layer; as it does so, suction becomes more and more dominant, always pulling water from wetter soil toward drier, whatever the direction.

The details of the suction process warrant a closer look (see figure 4.4). To keep the argument simple, suppose the rain has fallen on bare ground and that the wetting front is moving down through fine sand; tiny sand grains are the only solid material present, and the water moves through the interstices among the grains. The water and the grains are attracted to each other by

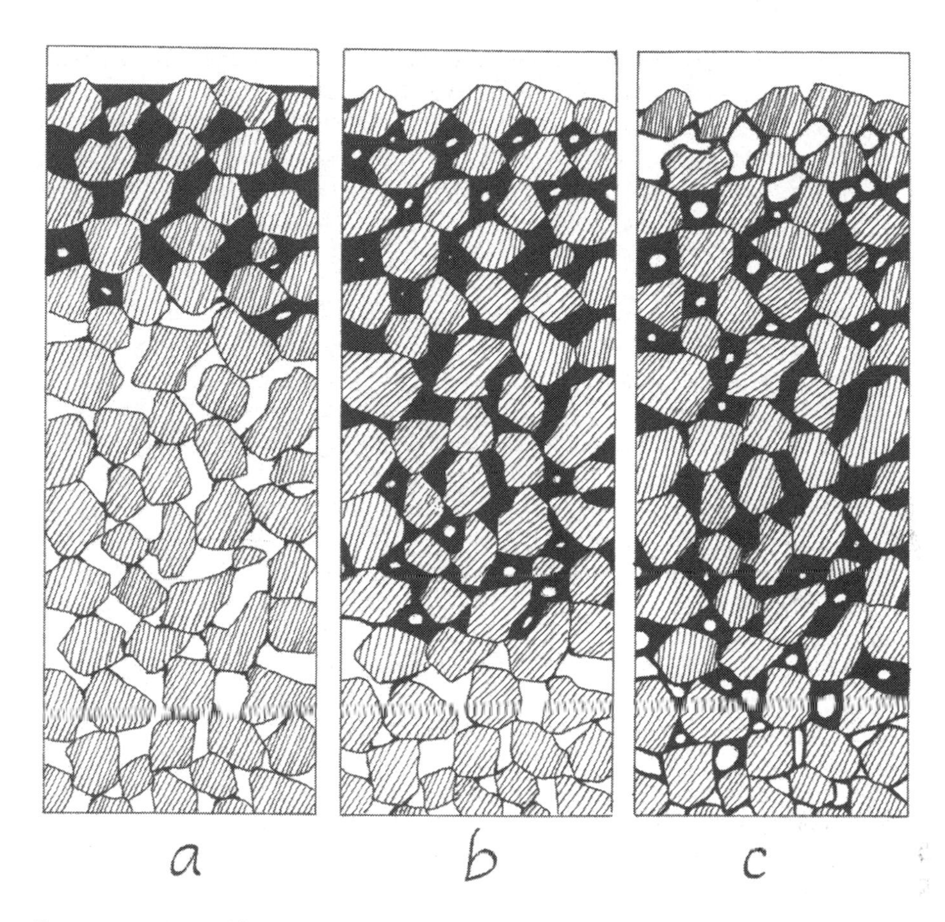

FIGURE 4.4. Water (black) descending through a vadose zone of fine dry sand after a heavy shower. *(a)* Newly fallen rain saturates the top layer; the wetting front is sharp. *(b)* Gravity and surface tension together pull the wetting front down; at the same time, surface tension holds water back at the trailing edge of the wet zone, which widens and becomes less distinct. *(c)* The wet zone sinks deeper, leaving the surface layer still damp; note the fragments of film remaining where sand grains touch each other.

molecular forces, causing the water to creep along the grains' solid surfaces and create thin films of water on them; in a word, the water wets the surfaces. The creeping movement cannot go on indefinitely, however, because it is counteracted by another molecular force, that of *surface tension,* which causes water molecules to cling to each other. This means that a water film cannot keep on getting thinner and thinner without end: the tension limits the possible thickness of a water film; in other words, it limits the distance a given amount of water can spread.

When gravity and surface tension balance each other, the water stops moving; it has wetted the surfaces of all the sand grains it can reach, and the wetting front has lost its distinctness. Where close-packed grains surround a minute gap, surface tension holds together enough water to fill the gap. A larger gap will be lined with a continuous water film, enclosing and trapping a pocket of air.

Now consider the soil left behind, above the descending moisture wave. Once it has had time to partially dry, the water film breaks apart; its remnants persist in the crevices where grains make contact. As drying continues, the remaining water retreats into ever tinier crevices with correspondingly tiny openings to the outside world; evaporation therefore becomes slower and slower, which explains why very fine soil material, especially clay, can go on drying for months or years and will still contain microscopic fragments of water film long after it has become "bone dry" in the ordinary sense. It also explains why water flows quickly through sand, and why sand dries fast: coarse sand grains may be more than a million times larger than silt grains, and the interstices among them are correspondingly large.

Now let's return to the descending front, and consider the speed at which it descends. Normally (but not invariably: see below), water flows faster through coarse-grained than through fine-grained material, as one would expect. In a dry climate, with the water table at great depth, the water from a rainstorm may take years to travel from the surface down to the water table if the soil consists of silt or clay. Even in otherwise sandy soil, a thin layer of clay or fine silt slows the flow of water tremendously. At least this is what usually happens, but not always.

Recall (section 2.3) that the speed of flow of groundwater depends on the force acting on it and on the hydraulic conductivity of the material it is flowing through. The same rule holds for water in the vadose zone, but there is a complication. The hydraulic conductivity of *un*saturated material depends on the amount of water already in it: water flows more slowly through dry material than through wet material. In particular, it will flow more slowly through *dry* sand than through *wet* silt. The surprising result is that a layer of sand can be a barrier to the downward flow of water through fine silt, if the silt happens to be wet and the sand dry.[7]

Complications such as these (and they are not the only ones) make the behavior of water in the vadose zone exceedingly hard to predict. But one way or another, sufficient water travels through the vadose zone to recharge the groundwater. On its way, at the bottom of the vadose zone's intermediate layer, it reaches the capillary fringe, mentioned already. The water in the capillary fringe is prevented from draining downward by surface tension. Groundwater creeps up the tiny channels in the subsoil in the way described earlier in this section: it stops creeping when the water drawn up into the channels becomes so heavy that surface tension cannot support any more against the pull of gravity. Then the water stops: in fine-grained subsoil the capillary fringe may be filled to saturation.

Now we return to the question raised at the end of section 4.1. Rephrased, it comes to this: given that groundwater and capillary water are not separated by any barrier, and that sometimes both kinds of water saturate the material holding them, what is the difference between them? The answer is a matter of water *tension,* or equivalently, negative pressure.

This is most easily explained by describing the workings of a *tensiometer* (see figure 4.5). The device is a narrow pipe, closed at the top and with a porous ceramic cup at the tip. Pipe and cup are filled with water and driven into the soil. A pressure gauge near the top of the pipe measures the pressure acting on the column of water. If the porous cup is in dry soil, water will be sucked out of the pipe through the cup's porous walls; that is, the water in the pipe will be under tension and its pressure negative. The pressure is negative whenever the tensiometer tip is in soil or subsoil that is not fully saturated with water and, also, everywhere in the capillary fringe, even if it *is* fully saturated. But at a certain level the tension disappears, and this level is the water table by definition. Indeed, the vadose zone is *defined* as the zone where the soil moisture is under tension.[8]

Tensiometers are familiar instruments to many people besides hydrologists. They are used by farmers, orchardists, truck crop growers and gardeners, as soil moisture probes.[9] Hydrologists use both tensiometers and piezometers (see section 2.4) to investigate subsurface water. They work on different principles (compare the two accounts) and are used at different depths: in general, piezometers are used to study groundwater and tensiometers to study vadose water.

So much for the way water moves in lifeless subsoil. Except in arid cli-

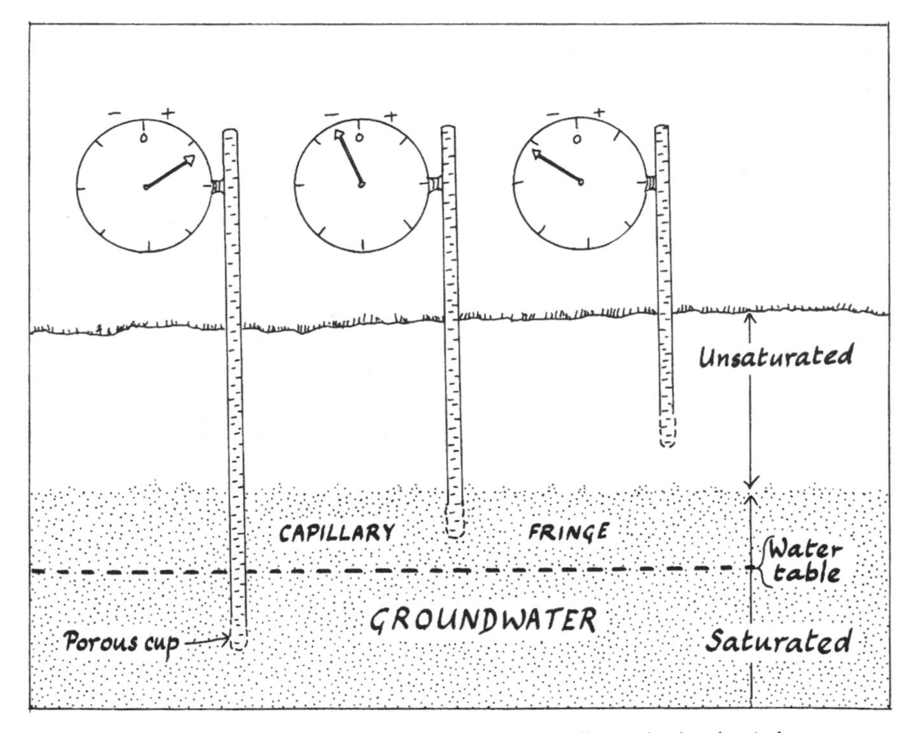

FIGURE 4.5. Three tensiometers driven into the ground to different depths; the circles represent their dials. As the diagram shows, pressure is positive when the tensiometer tip penetrates the groundwater; it is negative when the tip reaches only into the vadose zone, either into the capillary fringe or into unsaturated soil.

mates, it moves down under the pull of gravity far more often than it moves up because of surface tension. In life-containing soil, water behaves very differently. Plants can lift water upward to a degree surface tension never could. The remarkable effects of vegetation on the water cycle concern us next.

4.5 *Water Movement in the Soil: The Power of Plants*

The uppermost layer of the vadose zone is the soil; it is the layer in which water comes and goes most rapidly, because of the influence of plants. Plant roots absorb enormous quantities of soil water and transport it upward.[10] Although a fraction of this water is used by the plants to build their own tissues,[11] a much larger part of the water absorbed passes right through the plants and is "exhaled" (technically, *transpired*), as water vapor, through myriads of microscopically small pores (*stomata*) on the leaf surfaces.

The amount of water transported from soil to atmosphere by vegetation is prodigious. A single Douglas-fir can transpire 100 liters in a summer day.[12] Transpiration doesn't account for all the water vapor rising from vegetated land, however. Besides transpired water, there is also evaporated water—the vapor rising from wet ground surfaces and wet vegetation; in technical terms it is water that was intercepted but failed to infiltrate the soil. When the sun comes out after a summer rainfall, evaporation may give off more water vapor than transpiration for a while: wet surfaces are everywhere, and wet leaves transpire less rapidly than dry ones. But once the surface wetness has gone, transpiration soon overtakes evaporation. In keeping track of the hydrological cycle, it is easier to think of the *total* rate at which water is lost to the atmosphere from vegetated ground than it is to treat evaporation and transpiration separately. That is why, for measurement purposes, the two are usually lumped together and called *evapotranspiration.*

In the hour following a heavy rainstorm, 10 tons of vapor can rise from every hectare of an evergreen forest. As high a rate as this cannot be kept up for long: as the intercepted rain evaporates, the tree canopies dry out, whereupon only the transpiration part of evapotranspiration continues. All the same, a forest can lose 50 tons per hectare in one day.

Instead of measuring evapotranspiration as the weight of water lost over an area, it is often convenient to convert the result to the equivalent *depth* of water on the same area; then gains and losses of water, from rain and from evapotranspiration respectively, are easy to compare. For example, the loss of 50 tons of water from a hectare of forest comes to the same thing as losing 5 millimeters of rain. But it is hardly a "loss"; the flow of water through a plant is as necessary to the plant's life as the flow of blood is to an animal's life.

Grasslands, as well as forests, are tremendous producers of evapotranspiration. But comparing forests with grassland, or even one forest with another, is difficult: the amount of water a tract of vegetation can "vaporize" in a year depends on which seasons are wet and which dry. The proportion of the year's annual precipitation that a hectare of vegetated ground can send back into the atmosphere is much greater if the precipitation comes in summer when plants are actively growing and transpiring, than if it comes in winter when plants are dormant.

Transpiration is not visible. Even when it's going on all around you, there is nothing to see; but it is easily felt. Anyone who walks from a treeless, sun-baked field into a cool, moist forest on a hot summer day is instantly aware of transpiration—aware through the sense of touch; the welcome coolness is due as much to the water vapor in the air as to the shade. It is a mistake to suppose that the air in a forest is damp because trees in some way "create" water. They don't. Far from creating water, every tree is helping itself to a below-ground supply. As mentioned above, some of what is absorbed is used as tissue-building material. The remainder—the transpired water—is no less important to a tree's welfare, even though it moves quickly up and out of the tree. It carries dissolved nutrients from the soil into the tree; large volumes of soil water must pass through a tree for it to obtain all the nutrients it needs. Also, a tree transpiring is, in effect, sweating; the resultant cooling protects the tree from harmful overheating in hot weather.

Nearly all land plants (some exceptions are mentioned below) obtain their water as vadose water, from the soil. So if plants are to survive, the soil water needs to be continually replenished, all through the growing season. Sometimes replenishment is delayed and plants wilt; their cells collapse, as a balloon does when it loses air. If water comes soon enough the plants recover; if not, they die.

Wilting happens more often in garden plants than in wild ones adapted to local conditions. In particular, plants adapted to regions with dry summers are unaffected by short spells of drought. Many garden plants are strongly affected, however; they often need water more frequently than nature provides it. All gardeners are familiar with the behavior of such plants as lettuce and tomato that wilt on a hot, summer afternoon and recover splendidly in the night without being watered (though they won't keep doing this for many days on end!). What has happened is that, during the night, transpiration stopped because the plants' stomata closed in the dark; this gave time for the dried-out soil around the roots to absorb replacement water that moved up from a greater depth.

As we have seen, water is moved in the soil by both physical and biological forces. Of the physical forces, gravity can only move it down, but surface tension can move it in any direction. When it comes to biological forces, plants can move water up and *out* of the soil by transpiration, as we have already seen; they can also move water from place to place *within* the soil. Some deep-rooted trees and shrubs, sagebrush and sugar maple, for example, raise soil water upward by a process known as *hydraulic lift*.[13] Through the tips of their deep roots, which reach down into permanently moist soil, such plants absorb water at night as well as by day; the water moves up inside these roots, and some of it is exuded into the soil at a higher level via branch roots which spread sideways into shallow, dry soil. In this way the plants cache a supply of water, within reach of their shallow roots, to last them through the heat of the following day.

Plants capable of raising deep water by means of hydraulic lift no doubt evolved the mechanism as a way of supplying their own needs. But other, shallow-rooted plants benefit too, if they grow close enough to a "water lifter" to steal from its cache; the theft is known as *water parasitism*. It presumably explains why you sometimes find a diverse collection of obviously flourishing ground plants growing below the spreading canopy of a large, solitary tree (figure 4.6). They are water parasites, thriving on the water, and its dissolved mineral nutrients, raised from the depths by the tree. Because of their need for light, water parasites do best below isolated trees.

Water enters the soil not only from above (rain and snow) and below (groundwater raised by surface tension or by hydraulic lift) but also, occasionally, from the side. For example, it happens when stream-bank soil that has been sucked dry by plant roots, itself sucks replacement water from the stream beside it, instead of from the deeper soil levels below.

The result can be a valley woodland in mountainous country where young streamside trees get their water from one source—the stream—while the apparently identical nonstreamside trees get theirs from another source—rain. The water from the different sources sometimes differs chemically. It can also differ because of subtle atomic distinctions between the hydrogen atoms in the water, and it is this difference that makes it possible to know from whence a tree's water supply is coming.[14] A minute fraction of all hydrogen consists of "heavy hydrogen," otherwise known as deuterium,[15] which, when combined with oxygen and "ordinary" hydrogen, makes heavy water. Consequently, a minute fraction of all water is heavy water. However, the size of the fraction is not the same everywhere. In particular, there is more heavy water in warm, lowland rain than in melting snow high in the mountains. The streams in mountain valleys are fed by such meltwater. Researchers working in valleys in the Wasatch Mountains of Utah sampled the sap of trees (maples and oaks) growing in the valleys. Considering only the young trees—those with shallow root systems—comparisons showed that the sap of streamside trees contained a significantly smaller fraction of heavy water than did the sap of nonstreamside trees. The conclusion follows: the water supply-

FIGURE 4.6. A deep-rooted tree with a rich community of water parasites around it.

ing the streamside trees comes, via the streams, from mountain snows, whereas the water supplying the nonstreamside trees is summer rain.

If the water table is at or above ground level, as it is in wetlands and open water, there is no vadose zone. Wetland plants like cattails and bulrushes are rooted in soil saturated with groundwater right up to the surface for many months of the year, and they are adapted to these conditions (see chapter 10). Plants that rely on groundwater for most or all of their water supply are known as *phreatophytes*.

Only a few land plants are phreatophytes. The best known are many species of poplars and willows. Huge poplars growing out of bone-dry sand are a common sight on the banks and floodplains of broad, lowland rivers and on midriver islands. Their roots reach down to groundwater, assuring them of a continuous water supply regardless of surface dryness. The difficult stage of a poplar's life is its first few weeks. The tree starts as a newly fallen seed, germinating in spring on a surface wetted by spring floods; the seedling's roots must then grow fast enough for their tips never to be left behind by the sinking water table, which starts to drop when the floods subside. All is well provided the roots can keep up; otherwise the seedling dies.

4.6 *The Wetness of Soil: Three Degrees of Wetness*

Imagine a period of drenching rain followed by a long dry spell, and consider how the soil moisture changes. When the rain first stops, the soil is completely sodden, but it immediately begins to lose water and goes on losing it as the dry period lengthens. No sudden changes are noticeable. All the same, two changes of state occur in the soil as it dries.

The first change comes when no more water can drain from the soil under the pull of gravity. Before this point is reached, water will drip from a shovelful of wet soil under its own weight. When the dripping finally stops—in a very fine-grained soil it sometimes continues, more and more slowly, for weeks—the soil has lost all its *gravitational water.*

The loss continues, however, because plants' roots absorb and remove water held in the soil's pores. This is the water held in place by the force of surface tension, which is known as *capillary water* (note that—confusingly—the same name is used for the water in the capillary fringe, which may be far down in the subsoil; we are here considering water held motionless in true soil, within reach of plants' roots).

The second change in the state of the soil water comes when all the capillary water has been extracted. At this point the "living" force exerted by plants' roots is no longer strong enough to counteract the "nonliving" molecular force causing water to adhere to the surfaces of the soil particles. This remaining water is called *hygroscopic water.*

This division of soil water into three kinds, with distinct "behaviors," applies to any soil, whether coarse or fine. Figure 4.7 illustrates these various quantities for a few contrasted types of soil. One other technical term to mention is *available water.* It means gravitational water plus capillary water,

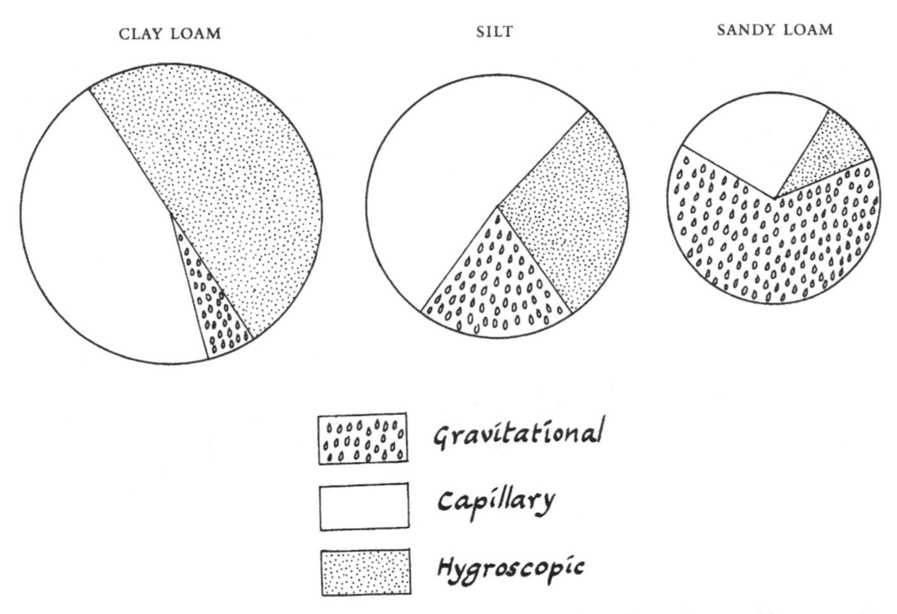

FIGURE 4.7. The division of soil water into three kinds (gravitational, capillary, and hygroscopic) in three typical soils (clay loam, silt, sandy loam). The areas of the circles are proportional to the porosities of the soils (the quantities of water they can hold).

which is all the water that plants can extract and use. Hygroscopic water is unavailable.

Further technicalities arise, because we are now in a branch of science where geologists, soil scientists, plant scientists, and agricultural experts (not to mention gardeners) are all apt to invent their own jargon without consulting each other. When soil contains its full quota of capillary water (and a little bit more; see below), it is said to be at *field capacity*.

Field capacity is not measured with any great precision. The usual procedure is this: 2 or 3 days are allowed to pass after a heavy rain has saturated the soil completely; a sample of soil is then taken, weighed immediately, dried in an oven, and weighed again. Subtracting the second weight from the first gives the amount of capillary water the soil had been holding plus any gravitational water that hasn't had time to drain. If the sample of soil has the shape of a cylinder, it is easy to find what the depth of an equal quantity of water would be in a container of the same cross-section. In this way, depth can be used as a measure of field capacity, making it easy to visualize field capacity as equivalent to precipitation of the same amount.[16] What makes the whole procedure so rough and ready, however, is that the wet soil sample is allowed to drain for only 2 or 3 days. This is too short a time for all the gravitational water to drain from any but the coarsest, sandiest soils; clayey soils may take weeks to drain. So field capacity overestimates the amount of capillary water a soil can hold, and the finer the soil the greater the error.

When a soil loses all its capillary water, because growing vegetation has extracted and transpired the last drop, it has reached the *wilting point.* More precisely, this point is reached when *permanent* wilting sets in; this happens when the nearest capillary water is so far below plants' root zones that there is no chance of moisture being restored in the night hours when transpiration stops. All that remains in the soil is hygroscopic water, unavailable to plants because it is held too tightly by the soil particles. The amount of hygroscopic water ranges from as low as 2 percent in sand up to 15 percent or more in clay.

4.7 *Surpluses and Deficits*

The rate at which vegetated land loses water by evapotranspiration depends on the amount of water in the soil and on the weather. As you would expect, for a given amount of soil moisture, evapotranspiration will be most rapid on hot, windy days, when the air is dry. But the rate is obviously limited by the amount of water available, however hot, dry, and windy the weather.

This has led scientists to differentiate between *actual evapotranspiration* (AE), and *potential evapotranspiration* (PE). As the names imply, the former describes evapotranspiration as it actually happens at a particular place and time, whereas the latter describes what *would* happen in the same weather conditions, given an unlimited supply of soil water, water provided by artificial irrigation, for example. Measurements of AE and PE are usually averaged over month-long intervals. Weather, too, is averaged, which means that the abrupt fluctuations of real weather are disregarded.

The numbers used to record evapotranspiration (both AE and PE) are depths, as explained in section 4.5. They represent the average depth of water lost, actually or potentially, at a particular observing station in a month. This makes them comparable with measurements of precipitation, recorded as the depth of water deposited at the same place in the same month.

Finding the AE and PE at a place for each month of the year entails some fairly complicated experiments and calculations.[17] The calculations are usually of the rule-of-thumb variety: that is, they use formulas based on past experience rather than equations derived from the laws of physics. The results are used in drawing up the average *water balance* (sometimes called *water* or *moisture budget*) of a district, an accounting of the rates at which water enters and leaves the soil, month by month through the year.

In months when the precipitation is greater than the PE, the water supply is more than enough to allow evapotranspiration to reach its full potential; that is, AE = PE. In addition, there is usually a moisture surplus. Some of the surplus soaks down into the ground to *recharge* the soil water, and the rest dribbles away on the surface as *overflow*.[18] Condensing this paragraph into a brief equation (the "wet-weather equation") we have

$$Precipitation = AE + Recharge + Overflow.$$

In months when the precipitation is less than the PE, the actual evapotranspiration (AE) must inevitably fall short of its potential. When this happens, there is a moisture deficit, equal to the amount by which AE falls short of PE. Plants use up all the incoming precipitation and make withdrawals from the soil's stored water in an attempt to cover the deficit. Sometimes they succeed; but in dry months, precipitation and withdrawals combined are still not enough to eliminate the deficit. This means that the "dry-weather equation" is

$$PE = Precipitation + Withdrawal + Deficit.$$

Note that the two equations are dissimilar. In the wet-weather equation all the terms relate to material "things": rain, water flowing into and over the ground, and vapor being evaporated and transpired. The dry-weather equation contains abstract notions: PE includes some evapotranspiration that doesn't actually happen, and the deficit is like an unpaid bill.

Figure 4.8 illustrates the two equations diagrammatically.[19] Note that the labeled bars on each side of the two drawings are of equal height; in each case, the bars represent the left and right sides of an equation.

Figure 4.8a illustrates the wet-weather equation. The height of the left bar represents precipitation, and that of the right bar represents AE and the surplus (recharge plus overflow) added together.

Figure 4.8b illustrates the dry-weather equation. The height of the left bar

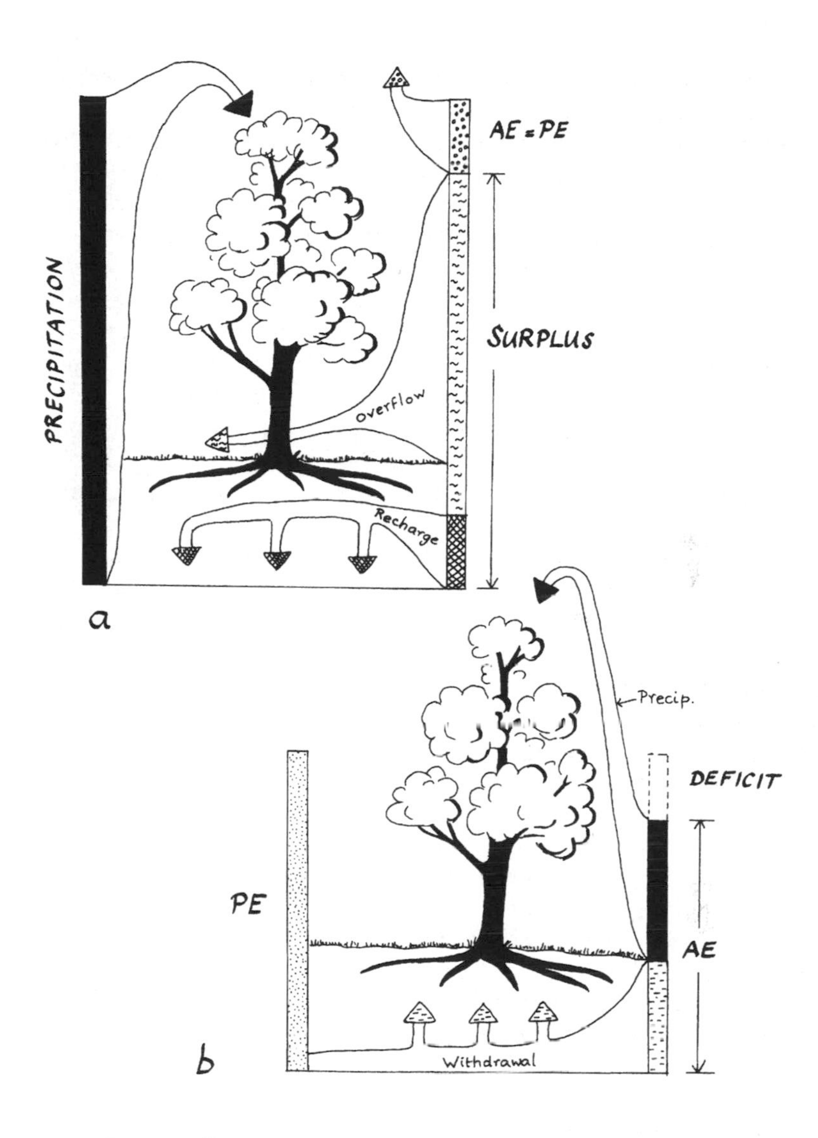

FIGURE 4.8. Diagrams illustrating the water-balance equations. The data relate to Vancouver, British Columbia, which has heavy rainfall in winter and drought in summer. *(a)* November. *Precipitation* (150 mm) = *AE* (20 mm) + *Surplus* (130 mm). Some of the November surplus recharges the soil moisture, but later in the winter, when the soil is saturated, all the surplus flows away on the surface. *(b)* June. *PE* (98 mm) = *AE* (78 mm) + *Deficit* (20 mm).

represents PE, and that of the right bar represents AE (precipitation plus withdrawal) and the deficit added together.

The overriding question is, what determines whether there will be a surplus or a deficit? One or the other must happen unless an unlikely coincidence occurs. The answer is that it depends on whether evapotranspiration is being limited by a lack of energy or by a lack of water.

In cool weather, or equivalently, if energy is in short supply, transpiration by plants is slow, as is the direct evaporation of water from moist surfaces. In these circumstances AE can "consume" only a small amount of water, often much less than that supplied by precipitation. Shortage of energy is the limiting factor.

Conversely, in warm weather, wet surfaces dry fast, and transpiration would be vigorous, given an adequate supply of water. In these circumstances it often happens that AE is held below PE by the lack of water. In other words, shortage of water is the limiting factor.

The worldwide distribution of different kinds of vegetation is controlled by regional variations in PE, which is, in a sense, a measure of energy, specifically of heat energy. For any place where the water balance has been worked out, it is easy to calculate the PE for the whole year simply by adding together the PEs for the separate months. This measures the amount of heat energy available to the vegetation during the year. Scientists have found that, provided water is ample (no deficits), the vegetation is tundra if the PE is less than 250 millimeters, coniferous forest if it falls between 250 and 600 millimeters, and deciduous forest if it exceeds 600 millimeters, well distributed through the year.[20] Note that for deciduous forests to thrive, warmth and ample water must come at the same time. Where PE exceeds 600 millimeters but summers are dry, the forests are coniferous rather than deciduous; these conditions—plentiful energy but a marked summer drought—account for the magnificent Douglas-fir forests of the northwest coast of North America. Conifers endure temporary drought better than deciduous trees, in part because their stiff needle-leaves are less damaged by a lack of water than are the softer leaves of deciduous trees.

To return from the global to the local scale, and from annual to daily—even hourly—changes. We have so far considered AEs and PEs averaged over big districts. These averages conceal notable local variations. For example, it often happens that AE is as high as PE in a valley that serves as a groundwater discharge area at the same time as AE falls far short of PE on the adjacent hill slope, which serves as a groundwater recharge area. As a general rule,

AE is equal to PE wherever and whenever the quantity of water in the soil equals or exceeds field capacity. Soon after you water a dry vegetable garden, the AE in the garden attains the PE level—in other words, the actual evapotranspiration becomes as high as it can be, given the ambient temperature—even though this is not true of the unwatered, adjacent land. Thoughts about AE and PE make watering the garden a mental as well as a physical exercise.

5

Flowing Water: Rivers and Streams

5.1 Water Flowing before Our Eyes

Underground water, the topic of the two preceding chapters, is rarely uppermost in the thoughts of anybody except a concerned professional such as a hydrologist, a hydrogeologist, or an engineer. Water at the earth's surface, by contrast, attracts the interest and attention of nearly everybody. Rivers, especially, affect the lives of millions, in direct and obvious ways. All over the world, rivers supply water for domestic, agricultural, and industrial use; navigable rivers are travel routes; dammed rivers generate hydroelectric power. One great river, the Ganges, is sacred to millions of Hindus. By periodically flooding, rivers can either restore the fertility of the soil (as in the delta of the Nile) or destroy the property and livelihood, sometimes even the lives, of thousands of people.

The material effects of rivers on people are only half the story, however. Equally important to most of us are their intangible contributions to our lives. Rivers and streams are an unfailing source of delight to anybody who enjoys the outdoor world. Many people are actively involved with rivers—boaters, anglers, and naturalists, for example. But even for those with no specific interest in them, the mere existence of rivers makes the world a more attractive and a more interesting place; without them we should be spiritually as well as materially deprived.

A vast amount of scientific information on the subject of rivers is available, and research scientists continue to enlarge the store of knowledge, for much remains to be learned. Knowledge about rivers isn't the private preserve of professional scientists, however. Anybody who keeps their eyes open, and makes deductions from what they see, can learn a considerable amount. Much of the information in the following sections can be checked by anybody, merely by observation and without using expensive equipment; still more can be gained simply from thinking about rivers and paying conscious attention to self-evident but seldom-contemplated facts. Indeed, if you were asked to write down all you know about rivers already, the amount would probably be much more than you realized. Becoming aware of one's own knowledge is one of the bonuses of paying attention to the natural world; observing rivers and streams, with their never-ending movement and change, is particularly rewarding.

5.2 How Do Streams Originate and Grow?

Most people, at least sometimes, have wanted to follow a stream or river upstream to its starting point. How far, and through what kind of country, has the water flowed before arriving at the point where you encountered it, and what would you find at its ultimate source? In Victorian times such questions motivated the travels of many of the world's great explorers; the search for the headwaters of the Nile, for instance, is one of the epics of the history of exploration. The mystery (from the European point of view) of the Nile's source was eventually solved by J. H. Speke in 1858, when he found that the river's main branch, the White Nile, flowed from Lake Victoria in East Africa. The St. Lawrence is another great river that flows from an enormous lake, in this case Lake Ontario.

To name a lake as the source of a river is, of course, an unsatisfying answer to the question, what is a river's ultimate origin? Big lakes from which rivers flow are always fed by smaller, inflowing rivers, and the sources of the feeder rivers are much harder to pin down. All rivers, or nearly all, however large they are by the time they flow into a lake or the sea, start in a small way, usually in hilly or mountainous country. (The cautious words "nearly all" are to exclude rivers that begin underground in limestone caverns, and rivers that begin as meltwater from the ends of glaciers.)

The majority of rivers begin at an indeterminate point in a slight depression in the ground, where groundwater happens to ooze out as a gentle seep; occasionally the water gushes from a true spring, but oozing seeps are far more common than gushing springs; indeed, they are very common, and usually overlooked. A depression also serves as a collector for overland flow from the surrounding slopes; overland flow is the rain that fails to soak into the soil after it reaches the ground. At first it flows as a widespread sheet of water on the surface, but it doesn't retain the form of a sheet for long; irregularities in the ground surface soon split it into rills. To hydrologists the word *rill* is a technical term meaning a miniature gully that the water from a single rainfall, or a single snowmelt, erodes for itself. Rills are formed anew on each occasion: they are not permanent features on the ground. Often they aren't even visible: in forested country the rills that carry away overland flow are concealed under fallen leaves and leaf mold.

Eventually seepage in the bottom of a depression, augmented by the water entering in rills from the surrounding slopes, adds up to enough water to erode a self-sustaining, permanent channel through which the water drains away: a stream is born. What makes a true stream, as distinct from water draining away over irregular ground, is its self-sustaining character; erosion by the stream deepens and widens its channel, ensuring that more surface water will be captured, which increases the stream's erosive power, and so on in a natural feedback.

This doesn't inevitably happen. If the depression in which water collects is a hollow with higher ground all round, then the accumulated water forms a pond, which can empty in several different ways: it may evaporate; it may soak down into the groundwater; it may overflow in a heavy rainstorm; or it may even drain away as an underground stream and emerge lower down beyond the hollow, as in figure 4.3.

When a stream is born, groundwater seepage is usually far more important than overland flow in bringing it into being. In general, only one-fifth of the water that reaches the surface as rain collects in streams and rivers.[1]

In suitable terrain—hilly or mountainous country, receiving ample precipitation—tiny streams form at innumerable scattered locations. As they flow downhill, they join with one another until a great many small streams have combined into a few large ones. The final result is a complex drainage system. The streams all flow down valleys, but since valleys are created by streams, we seem to have reached a chicken-and-egg situation: which comes first, a stream or its valley?

If it were possible to follow a stream back in time, as well as back upstream to its source, one would most often find that the birth of the stream initiated the formation of its valley; as outlined above, the first trickles of water eroded an incipient channel, which then guided later flows; the later flows progressively widened and deepened the channel until it became a valley with the water confined to the bottom of it. Only in special circumstances can the whole sequence of events be followed by one observer; this can be done when a volcanic eruption, like that of Mount St. Helens, Washington, in 1980, creates an entirely new, uneroded ground surface for groundwater to seep through and rain to fall upon, or when a reservoir is drained suddenly and its bed provides a similarly new, uneroded surface.

Drainage systems change and develop continuously, over centuries and millennia, as geological forces cause landmasses to rise and fall, or tilt, or bend, with customary geological slowness. These gradual topographic changes sometimes create new valleys, initially dry, that subsequently become the channels of new streams and rivers. When this happens, the valley appears first and its stream comes later.

To return to the present: consider what happens downstream from a stream's source, the place where (as we saw above) the combined flow from seeps and rills develops a current strong enough to erode a channel, and after that a valley big enough to persist. Rain draining down the valley's slopes or falling on the stream itself, and groundwater seeps developing on the slopes and in the streambed steadily increase the volume of flowing water; by degrees, the stream becomes a river. (These words are ill-defined; everybody has their own opinion on how large a stream must be to be called a river.) Small streams combine to form larger ones; when the combining streams are markedly unequal in size, the smaller is described as a tributary of the larger.

The combined effects of rainfall, groundwater seepage, and the arrival of tributaries cause most rivers to grow progressively larger. The exceptions are small rivers that flow from mountains into desert; with insufficient rain to sustain them, and the groundwater too deep to contribute, they evaporate or soak into the ground: in short, they run dry.

Indeed, two different kinds of streams, or parts of streams, can be distinguished, *gaining streams* and *losing streams*.[2] The former gain groundwater from submerged seeps and springs, whereas the latter lose some of their water, which soaks away to recharge the groundwater. Except in arid climates, gaining streams are the commoner of the two. The groundwater in a gaining stream is known as the stream's *baseflow*.

5.3 Watersheds

When you trace on a map the course of a major river from its tiny beginnings seaward, it is obvious that its flow becomes larger and larger. Every tributary brings a sudden increase, and between tributaries the river grows gradually, because of overland flow entering it directly.

It is not only the river that grows, however. Besides the river itself, its whole *watershed* or *drainage basin*—the land that drains into it—grows too, in the sense that it embraces an ever larger area. Every time a tributary stream joins the river, the tributary's own watershed is added to the river's previous watershed. As tributary watersheds become part of an ever more inclusive watershed, one after another as you go downstream, the area of land draining into the "master" river grows larger in abrupt "jumps" (figure 5.1).[3] In this way, the watersheds of big rivers expand to enormous size; the area of land draining into the Mississippi by the time it reaches New Orleans is almost 32 million square kilometers, for example.

A digression on words is necessary here: in North America the terms *watershed* and *drainage basin,* as defined above, are used as synonyms, and *watershed* is the usual nontechnical word. The height of land dividing adjacent watersheds is known as a *drainage divide,* or simply a *divide.* In Britain and Europe, by contrast, the word *watershed* is synonymous with *divide.* The easiest way to avoid misunderstandings would obviously be to dispense with the ambiguous word *watershed* altogether, and speak only of *drainage basins* and *divides.* Unfortunately, this solution is probably unattainable, now that *watershed* is so widely used. It is used in the North American sense in this book.

A river's behavior is greatly affected by the characteristics of its watershed. For instance, the steadiness of the river's flow at a given point is controlled by the area of the watershed upstream of the point. In a big watershed made up of numerous tributary watersheds, heavy rains in one of them will often be offset by a dry spell in another; in this way, local fluctuations in water supply become averaged out, making for a more even flow downstream.

Watersheds differ from one another in many features besides area. Much depends on climate, topography, and geology. One of the things an attentive traveler can't help noticing is the *drainage density.* This is the average length of the stream channels per unit area of land within the watershed.[4] Different kinds of landscapes have different drainage densities. For instance, the drainage density would be low in a gently rolling, densely forested landscape on

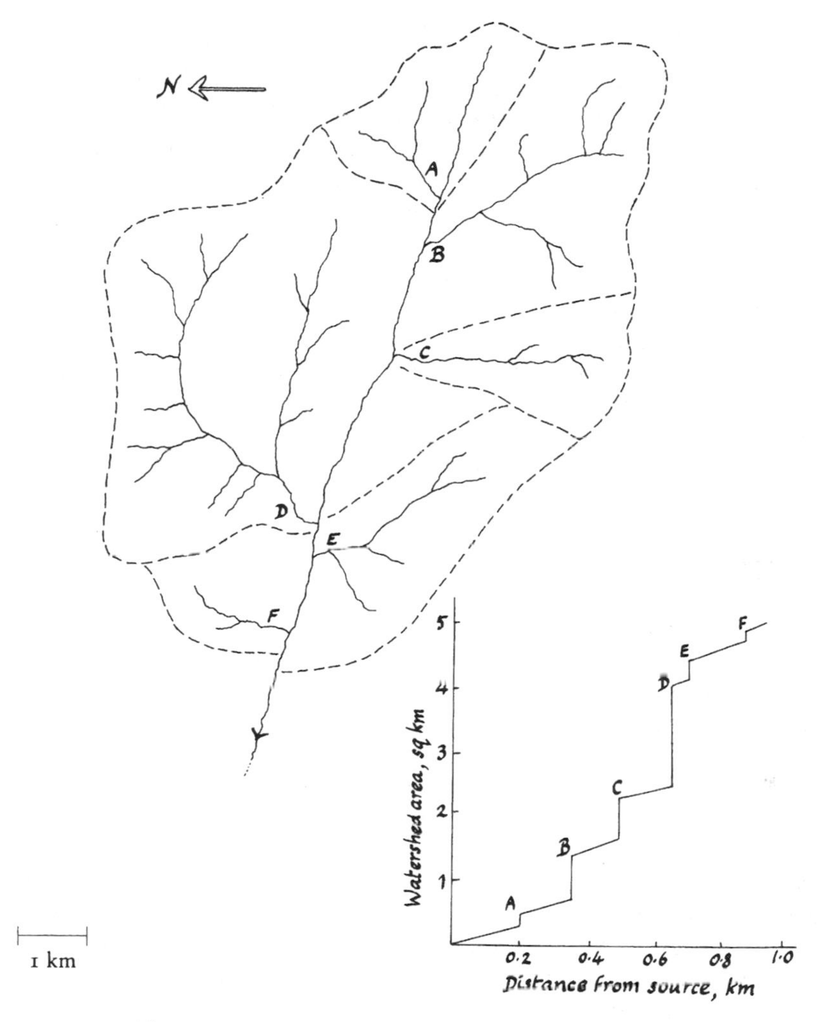

FIGURE 5.1. The upper watershed of Ursus Creek, Clayoquot Sound, British Columbia. Dashed lines show the divides. The tributaries are labeled with capital letters (A to F) where they join the main creek. *Inset:* Graph showing how the total watershed area (square kilometers) increases with increasing distance (kilometers) downstream from the source of Ursus Creek. Note the stepwise increase in watershed area as each tributary joins the main stream; each tributary's letter is shown at the top of its step.

massive, ancient granites. The bedrock would not be easily eroded by streams; and because so much of the rainfall would be taken up and transpired by the trees, comparatively little water would flow on the surface. In such a landscape there might be only 2 or 3 kilometers of channel per square kilometer of land. By contrast, consider a sparsely vegetated region where the surface is easily erodible, weakly consolidated clay; the drainage density might be 20 or 30 kilometers of channel per square kilometer, or ten times as great as in the preceding example. In arid, badlands terrain it can be ten times as great again, that is, 200 or 300 kilometers per square kilometer;[5] the tiny valleys are dry most of the time but carry sizable streams after summer thunderstorms.

Another characteristic of a watershed is its annual *runoff*. This is the amount of water leaving the watershed in a year. The bulk of it comes from overland flow and from rainwater that seeped or flowed for some distance in the vadose zone before entering the river. The rest of the water is baseflow (groundwater discharging into the river's channel), plus water from any lakes that chance to be part of the river's course; lake water is water in temporary storage. But whatever its immediate origin, a river's ultimate source was rain or snow that fell on the watershed.[6] Therefore, provided the quantity of water held in storage in a watershed doesn't change, the quantity of water a river receives that does *not* flow out in the river's channel must be lost by evapotranspiration within the watershed. As a simple formula,

Runoff = Precipitation − Evapotranspiration.

The way in which runoff is measured depends on the objectives of the person seeking the information. An engineer usually wants to know how much water flows out of a whole watershed in a chosen unit of time. The answer might be in metric tons per year. Here are two examples: the flow from the Mississippi is about 580 billion metric tons per year;[7] from the Fraser River in British Columbia, it is nearly 100 billion metric tons per year.[8]

Somebody wanting to judge how climate and vegetation affect a region's water budget would want to know the runoff per unit time *per unit area* of the watershed, and would be concerned with watersheds small enough for conditions to be much the same throughout. Dividing the runoff volume by the area it has come from gives the depth of an equal volume of water spread over a flat area of the same extent. Recall (section 4.5) that evapotranspiration is measured in the same way. Imagine a forested watershed of 100 square kilometers, in a region where the precipitation is 1,200 millimeters per year, and suppose that the runoff per year is 36 million cubic meters. The depth of this water if it were spread evenly over 100 square kilometers would be 360 millimeters; this is a convenient measure of runoff in the context, and it can also be plugged into the formula given above. The formula, rewritten as

$$Evapotranspiration = Precipitation - Runoff,$$

is often used to estimate the evapotranspiration from a watershed. Thus in the watershed we visualized, the evapotranspiration is 1,200 mm − 360 mm = 840 mm. This is a useful method of estimating evapotranspiration, which is difficult to measure directly.

5.4 A River's Flow

Measuring the runoff of a big river is a big undertaking, and obviously far beyond the scope of amateurs. But measuring the *discharge* of a small stream is a simple matter. A stream's discharge at any chosen point is the volume of water flowing past the point in a given time interval, usually a second. Measuring it requires no special expertise, no gadgets beyond a stopwatch, and no more than two or three people.

The procedure is what common sense suggests. All that needs to be known are the cross-sectional area of the flowing water and its speed of flow; the calculations are simple arithmetic.

Here is an outline of how to do it, assuming the stream is safe to wade. A note on judging whether a stream *is* safe to wade is given after the list.

CHOOSE a straight stretch of the stream, with clear, uncluttered banks.

STRETCH a rope across the stream and fix both ends, by tying them around trees, for example. The rope should have marking tabs, or spots of paint, at equal intervals. The length to choose for the interval is a matter of judgment; it depends on the width of the stream and the irregularity of its bed. Consider an interval of about one-fifth of the stream's width, and modify this as conditions suggest.

WADE IN and measure the depth of the stream below each of the tabs on the bridging rope. This can be done with a meter rule, a surveyor's staff, or a measuring tape sufficiently weighted for it to hang vertically despite the current. The measurements are used to draw a cross-sectional profile of the stream from which to estimate its cross-sectional area, as shown in figure 5.2.

CHOOSE a length of the stream along which to measure the current velocity. Mark each end of the chosen stretch with a peg or distinctive rock on the bank; the stretch should bracket the point where the cross-sectional profile was measured. As to the length of the stretch, two or three times the width of the channel is usually sufficient.

NOW WADE IN and measure the current velocity below each tab on the bridging rope. Each measurement is made by putting a float into the stream and timing it, with a stopwatch, as it is carried from the upstream to the downstream marker. The float should start its journey a short way upstream of the upstream marker so that it is up to speed by the time it enters the measured stretch. Any object that floats without riding too high in the water, where air resistance would slow it or wind deflect it, is suitable. An orange is ideal (and traditional) for the purpose; some people eat the orange first and then use the peel. Orange or orange peel has the advantage of being easy to see, but unlike a stick or other piece of natural flotsam, it is "garbage" that must be retrieved.

AVERAGE the velocities and multiply the result by 0.85. This number corrects for the fact that the velocity below each tab was measured only at the surface, whereas ideally it should have been measured at a number of different depths and these results averaged too; the flow is faster close to the surface than at deeper levels.[9]

MULTIPLY the corrected average velocity by the cross-sectional area, and the answer is the stream's discharge. The most convenient units for measuring discharge are cubic meters per second, abbreviated to c.m.s. Figure 5.2 illustrates the calculations.

To judge whether flowing water is safe to wade, multiply its depth in meters by the speed of flow in meters per second (m/s); then avoid wading without a life jacket if the result is greater than one.[10] Since you cannot foretell the depth of the water ahead of you, apply the test repeatedly as you wade. To make this rule of thumb even easier to use, remember the following: current can be disregarded if the flow is gentle, say less than 1 m/s, about the speed of dawdling as opposed to purposeful walking; in a swift stream, flowing faster than a brisk walking speed (about 2 m/s), avoid wading into water deeper than 50 centimeters. Few streams flow faster than 3 m/s (nearly

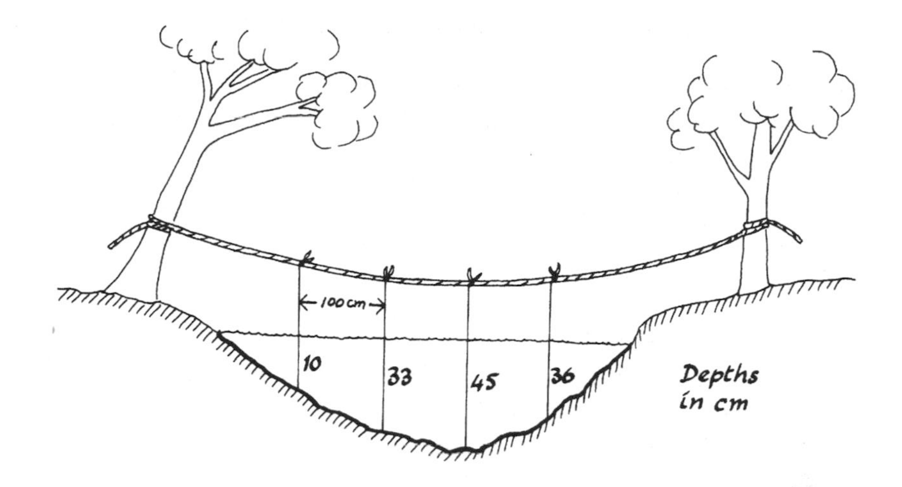

WIDTH OF STREAM = 5 meters

AVGE DEPTH = $\frac{1}{4}$ (10 + 33 + 45 + 36) cm = 31 cm

AREA OF CROSS-SECTION (APPROX) = 5 × 0·31 sq m = 1·55 sq m

SURFACE VELOCITIES BELOW THE FOUR TABS (METERS PER SECOND)

0·2, 0·7, 1·0, 0·5 m/s

AVGE SURFACE VELOCITY 0·6 m/s

ASSUMED AVGE VELOCITY OVER WHOLE CROSS-SECTION =

0·6 × 0·85 m/s = 0·51 m/s

APPROXIMATE DISCHARGE = 1·55 × 0·51 c.m.s. = 0·79 c.m.s.

FIGURE 5.2. How a stream's discharge is estimated. The length of the "wet edge" of the cross-section (shown with a heavy line) is known as the *wetted perimeter*.

11 kilometer per hour), for which the maximum safe depth would be around 30 centimeters.

It only remains to mention that the "floating orange" method of measuring stream currents is, of course, exceedingly rough and ready. Professional hydrologists use current meters; one type has a propeller mounted on a horizontal axis; another type resembles an anemometer (wind-speed meter), with cups at the end of spokes radiating from a vertical axis.

5.5 Rising and Falling Water Levels

Anybody who has observed a favorite stream or river through the seasons is aware that its water level varies through the year, being high in the rainiest months and lower in dry periods. Its discharge obviously varies correspondingly: when the level is high, the volume of water in the channel is great and the flow swift; the opposite holds when the river is shallow.

This gives a shortcut method for judging the discharge of a river at a particular point. All that need be known is the relationship between water level and discharge at the point; then the discharge can be inferred simply by measuring the water level. Hydrologists acquire the needed data by making accurate measurements of the discharge on several occasions, with the water at a different level each time. The results are graphed as a *rating curve* (figure 5.3).

The water level, or *stage* as it is called, is the height of the water surface above some arbitrary, chosen zero level. It doesn't matter what this standard is, provided it is definite and fixed. The simplest way of measuring a river's stage is with a *staff gage,* the graduated ruler often to be seen fixed to a post or a bridge pier.

A rating curve shows the relation between stage and discharge only for the *gaging station* where it was made. Every gaging station has its own, unique rating curve. Observing the water level on a gaging staff is the most direct (and probably the most dependable!) way of measuring a river's stage, but less primitive methods are common nowadays. Many gaging stations have water-level recorders that run continuously. The equipment is housed in a small riverside cabin (that could be mistaken for a badly sited outhouse) and consists of a well connected to the river so that the water level is the same in both; the up and down movements of a float on the surface of the well

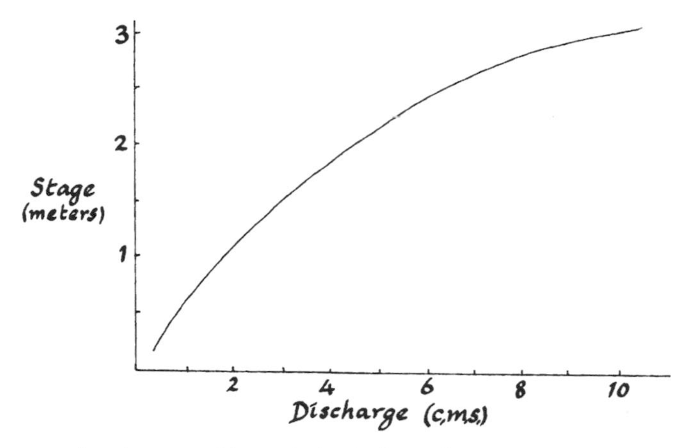

FIGURE 5.3. A rating curve. The vertical axis shows the stage in meters, as measured on a staff gage fixed to a bridge; the corresponding discharge in cubic meters per second (c.m.s.) is read off the horizontal axis.

water are recorded as ink lines on a chart fixed to a revolving drum, or, in more up-to-date versions, the float activates a digital recorder. In another type of equipment, the water pressure at the bottom of the well is measured with a pressure gauge; the height of the water above the gauge can be calculated from the pressure.

It is even possible to discover the stage of a big river by remote sensing from a satellite;[11] this method cannot record a river's stage continuously, but only at the instants when the satellite, in its orbit, is passing overhead.

The data from a gaging station—or better still, from a series of gaging stations along a river's course—enable a hydrologist to keep track of the way the river behaves over the course of time.

The data are plotted as a *streamflow hydrograph;* figure 5.4 shows an example. The hydrograph in the figure shows the behavior of a stream over a 5-week interval with a showery period near the beginning. The stream concerned is *perennial;* that is, it flows all year without drying up completely even in prolonged dry periods. What enables it to do this is an unfailing supply of stored water, which may be groundwater seeping in, or the flow from any permanent lakes that happen to be upstream. Indeed, a perennial stream's

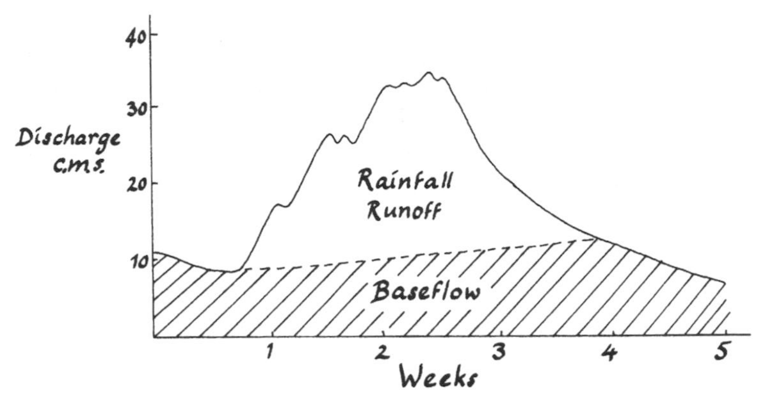

FIGURE 5.4. A 5-week section of a river hydrograph. The flow assumed to be baseflow is shaded. Repeated showers, starting near the end of week 1 and ending halfway through week 2, yielded the rainfall runoff, which combined with baseflow to give the total discharge.

discharge has two independent ingredients: that from stored water, which usually flows steadily with only gradual, minor fluctuations, if any; and runoff, also called *quickflow,* which varies from day to day—sometimes from hour to hour or even faster—in response to rain or melting snow. If there are no lakes along a river's course, the flow from storage consists of baseflow.

A hydrograph shows the total discharge of a river: nothing in the hydrograph distinguishes baseflow from runoff, and the two kinds of water are, of course, mixed and inseparable in the river itself. All the same, hydrologists attempt to make a distinction between flow from the two sources, and to differentiate between them on the finished graph, as shown in figure 5.4. Unfortunately, the methods of separating them are based on arbitrary and inconsistent rules of thumb,[12] or else on subjective judgment, which may be just as good.

Not all streams have a baseflow component, or at least not a permanent one. If they never receive baseflow, equivalently if they are *losing* streams at all times, they are called *ephemeral;* the flow of ephemeral streams consists wholly of runoff, so they flow only while it is raining and for a short time afterwards. Intermediate between *perennial* and *ephemeral* streams are *intermittent* ones; these are gaining streams in the wet season, but losing streams the rest of the time.

Another way of describing the three kinds of streams is by noting the behavior of the water table. The bed of a perennial stream is, at least in places, below the level of the water table at all seasons. Conversely, the bed of an ephemeral stream is above the water table at all seasons. Intermittent streams flow in regions with a marked seasonal contrast in precipitation, causing the water table to shift up and down correspondingly; the stream flows steadily when the water table is above the level of the streambed, at least here and there; but it becomes a losing stream in the dry season, when the water table sinks below the streambed.

Hydrographs modified to show only runoff, with the baseflow component deleted, are often drawn to illustrate how rainstorms of different intensities, and in different parts of a big watershed, affect a river's flow. They are useful for predicting floods and are called *storm* (or *flood*) *hydrographs.*

Not surprisingly, a stream is most likely to flood if it responds quickly to a nearby rainstorm; if storm water flows into a stream faster than it can be carried away, any stream must overflow its banks. This happens most often in arid country with steep, unvegetated slopes where overland flow is unhindered. A storm hydrograph for such a stream shows the rapid arrival and departure of storm rain as a high, narrow peak. A sudden intense storm, directly over a small watershed, will produce an especially spiky hydrograph at a gaging station just downstream of the storm center.

A stream in level, well-vegetated country is much less likely to flood. Vegetation slows overland flow, and rainwater has time to soak into the soil and be used up by evapotranspiration. Thus, less water from a storm reaches the stream, and the water that does arrive comes in by slow degrees. A storm hydrograph from such a stream shows the arrival and departure of storm rain as a low, wide hump. If the runoff comes from a gentle rainfall, or if it forms only part of the runoff from a large watershed whose entire output is monitored at the gaging station, the hydrograph will be smooth and undramatic.

The time it takes for storm flow to reach its peak varies too. The *lag time,* the delay between the midpoint[13] of the storm and peak streamflow, is much less in "flashy" streams (those with a sharply peaked hydrograph) than in sluggish ones (with a gently humped hydrograph). And the shorter the lag time, the more quickly will the runoff from a rainstorm flow away, leaving only baseflow in the stream's channel. From start to finish, the runoff from a storm normally takes about 3.5 lag times to flow past a gaging station.[14]

The shape of a storm hydrograph doesn't depend only on the storm and the kind of watershed affected by it. It also depends on how far downstream of the storm the hydrograph was made. Figure 5.5 shows three hydrographs of the same rainstorm, made at a series of gaging stations at increasing distances downstream from the storm center.

The first sign of storm runoff appears on the upstream hydrograph soon after the rain starts to fall; it appears later and later as you go downstream, as does the moment of peak runoff. At the same time the shape of the peak becomes progressively lower, wider, and less sharp, and it takes longer and longer for all signs of the storm to disappear.

5.6 How Fast Can a River Flow?

As anybody can easily observe, a river flows faster when it is in flood than when its level is low: the greater the volume, the greater the speed. The fact is so familiar that it is seldom pondered. But what *does* control the speed at which a river flows? The problem deserves some attention.

The way in which the speed of flow changes as water level rises and falls is discovered from observations made at a gaging station. Note that we are here concerned with the speed of flow averaged over the river's whole cross-section, disregarding for the moment the way the speed varies with depth and distance from the banks.

Observations have shown that the chief factors controlling speed are these: the cross-sectional area of the moving water, which is larger when the water level is high than when it is low; the slope of the river (a river has to slope if it is to flow, but the slope is often too slight to be perceptible without precise surveying instruments); the roughness of the river's bed and banks; and the length of the boundary between flowing water and motionless ground, known as the *wetted perimeter* (figure 5.2). As would be expected, the flow will be fast if the river's cross-sectional area is large or its slope steep. Conversely, it will be slow if the channel walls are rough or if the wetted perimeter is long.

Formulas have been devised to allow a river's speed to be calculated from these four factors. The subject is hardly new: the two classic formulas, still in use with modern refinements, are Chézy's, devised by a French engineer in 1768, and Manning's, devised by an Irish engineer in 1889.[15]

One use of the formulas is to estimate what the maximum discharge must have been the last time a river was in flood, after the flood is over and the opportunity for measurement is gone. The peak discharge can be estimated

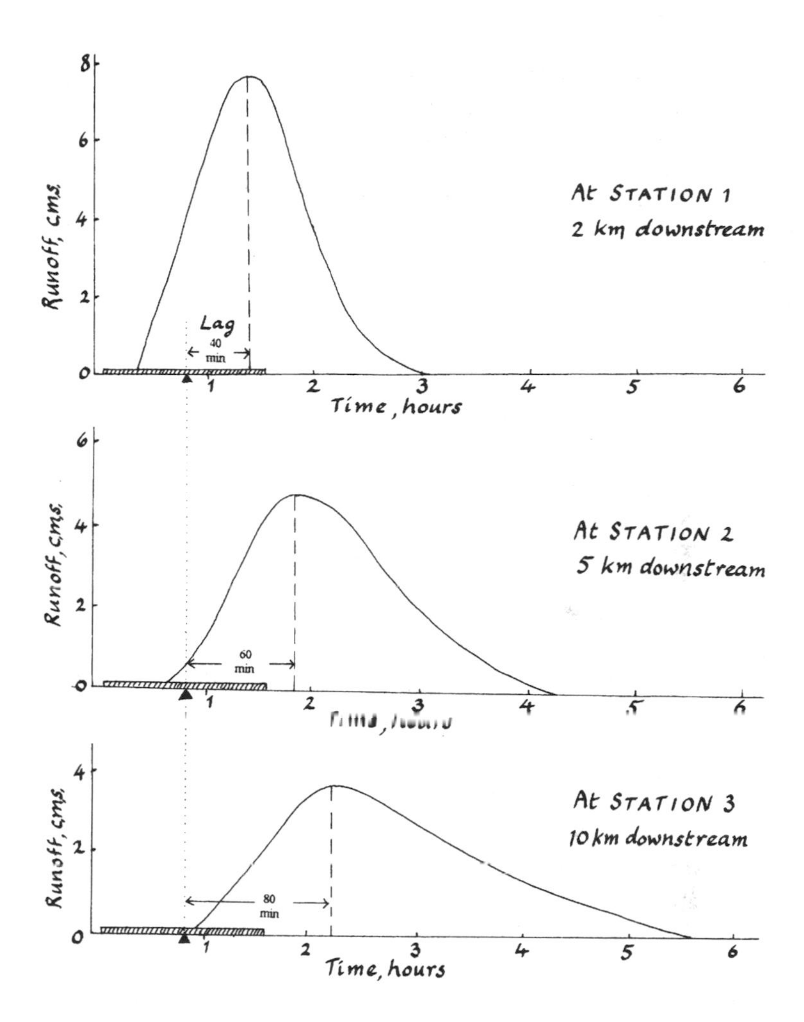

FIGURE 5.5. Storm hydrographs recording the same rainstorm at three different gaging stations; station 1 was nearest to the storm center, stations 2 and 3 successively farther downstream. The Y-axis (vertical) measures runoff, found by subtracting the estimated baseflow from the total flow. The X-axis (horizontal) measures time. The 1½-hour rainstorm that caused the runoff is shown by the shaded bar near the left end of each X-axis with an arrowhead marking the midpoint, and is the same in each graph. The lag times are shown. Note that the runoff first appeared at the successive stations 18 minutes, 34 minutes, and 55 minutes, respectively, after the beginning of the storm.

after the event, provided some bits of flotsam remain on the banks to mark the level of high water; the flotsam makes it possible to discover what the wetted perimeter and the cross-sectional area were when water level was at its highest. These two measurements, plus a knowledge of the channel's slope and the roughness of its bed, can be plugged into the formulas to give estimates of the peak velocity and the peak discharge.

The formulas are based more on rule of thumb than on the laws of physics and were designed more to predict the hydraulics of water in artificial canals than in natural streams. Moreover, they apply only to *uniform flow,* that is, to flow whose depth and velocity stay the same over considerable distances, which is seldom true of small streams with their alternating shallows and deeps, though it is approximately true of big rivers flowing through flat country. The formulas are said to work poorly for streams with slopes of more than 1 in 100; expert opinion varies enormously on how the roughness of a stream channel is best measured; and when applied to natural streams, the formulas may result in errors of more than 50 percent.[16] In spite of all these drawbacks, the formulas are still better than nothing.

Recall that the formulas can be applied only to a river in uniform flow; this means that its speed is not accelerating, which implies that the velocity is a *terminal velocity.* The concept of terminal velocity is commonplace in the context of objects falling through the air. However far it falls, no object keeps on accelerating indefinitely; once it has reached a speed at which gravity and air resistance balance, it maintains that speed until it hits the ground. The terminal velocity of a falling human being, for example, is about 200 kilometers per hour (km/h).[17]

The same arguments apply to flowing water. When a river's rate is unvarying, it is flowing at the terminal velocity appropriate to its water level at the moment. If the water level drops, the flow will slow down until it reaches, and remains at, a new, lower terminal velocity. Conversely, if the water level rises, the flow will speed up until it reaches a higher terminal velocity.

It follows from all this that a river in uniform flow that is full to the brim, with water to the top of its banks, is flowing at the maximum terminal velocity it can attain, the terminal velocity corresponding to maximum discharge. Many observations have been made to discover how fast rivers actually flow when in flood.[18] For rivers of average size, the maximum is usually less than 5 km/h, but it can be as much as 10 or 11 km/h in large rivers. A fast river entering a narrow gorge, where the funneling effect speeds the flow tremendously, can reach a velocity of 25 to 30 km/h, but when this happens the flow is not uniform and the classic formulas don't apply. The topmost speed of water falling in a waterfall is about 20 km/h.[19]

5.7 *Varying Currents and Their Effects*

Up to this point we have been concerned with a river's rate of discharge and consequently with its *average* velocity. Now we consider how the flowing water actually behaves in detail: instead of thinking of it as if it acted like toothpaste coming out of a tube, we treat it more realistically, as consisting of individual "threads" of water continually sliding past each other.

It is easy to see that this is what happens, simply by watching floating leaves drifting down a stream. They move more slowly close to the banks than near the center and submerged, waterlogged leaves, floating just off the bottom, are left behind by those above them. The result is that the water becomes more and more spread out lengthwise. Another line of argument leads to the same conclusion. Recall figure 5.5; it shows how a whole storm's worth of water progresses downstream as a unit, from one gaging station to another, neither gaining nor losing in volume on the way.[20] The hump in the hydrograph becomes progressively lower and wider, showing how the stormwater becomes spread through a longer and longer stretch of river.

To return to the details. The way in which a stream's current varies at different points in a cross-section is best shown in the form of a "contour" map, as in figure 5.6. The contours, known as *isovels*, are lines joining points

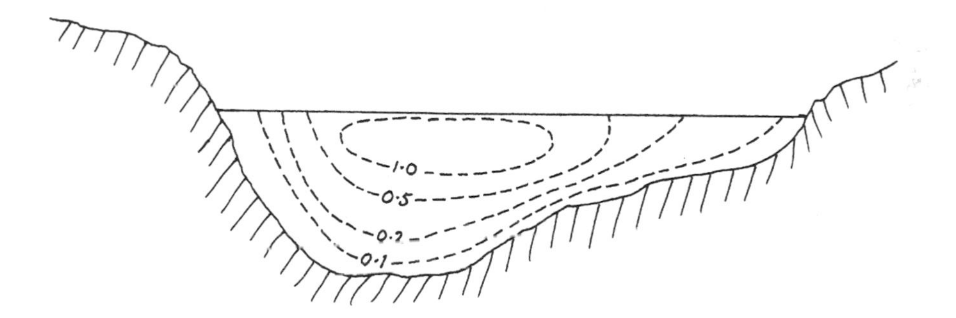

FIGURE 5.6. Current velocities at different points in a river. The isovels (dashed lines) join points of equal velocity. The velocities, in meters per second, are shown. Note that the flow is most rapid about halfway between the banks, just below the surface of the water.

of equal downstream current velocity. The example is typical; it shows how the current slows toward the sides and bottom of the channel and also (usually but not invariably) toward the water surface. This slowing at the top is surprising: more on the subject below. The slowing close to the solid walls of the channel—its bed and banks—is unsurprising (at least at first thought, but see section 5.9); the walls, especially if they are rough, would be expected to resist the water's flow, and, in fact, they do.

Note that the velocities in an isovel diagram are all strictly downstream velocities; they have been measured parallel to the river's banks or, equivalently, at right angles to the cross-section. They are the dominant element in streamflow, and comparatively easy to measure. However, as you can easily see by observing a stream, its flow is seldom directly downstream all the way across; crosscurrents often appear close to the banks, as shown by the movements of floating fragments of twigs and leaves. The crosscurrents may be rather erratic at the surface, where they are visible, and give only a hint of what is going on below. Careful studies of the details of streamflow have shown that in straight stretches two helical currents are apt to develop side by side,[21] as shown in figure 5.7. The water tends to flow from the banks toward the center at the surface, to "downwell" at the center, and then flow back toward the banks at the bottom.

Two consequences of these helical currents are, first, that a river's surface is often arched, being higher at the center than close to the banks; the curvature is very slight and seldom detectable without surveying instruments. The second consequence is easily seen when a stream is in spate (provided two side-by-side circulation cells have indeed developed, which doesn't invariably happen); floating particles and groups of bubbles at the surface are carried by the converging currents to the center of the river, and collect there to form a distinct lane of flotsam; the water is clear of floating material between this lane and the banks.

Now recall figure 5.6, showing the downstream current as slightly slower at the surface than just below it. The surface slowing is often attributed to friction between the moving water and the air above, but this isn't necessarily the true explanation.[22] Probably a more important cause is the helical currents; they give the movement of the water at the surface two components, in different directions; one component (the chief one) is downstream, the other component is across the stream. The total flow may be no slower at the surface than below it, but a current meter responding only to the downstream component underestimates the total velocity.

FIGURE 5.7. The flow (away from the viewer) of the water in a straight stretch of a river. The helical dashed lines show current directions within the river. Note the narrow lane of flotsam left floating in the middle of the river where the converging currents descend.

5.8 The Rough and the Smooth: Choppiness and Hydraulic Jumps

One of the chief differences between a big, majestic, lowland river and a rocky, highland stream is that the big river flows for long distances at the same unvarying speed, whereas the flow of the rocky stream switches from slow to fast and back again repeatedly, being swift over rock outcrops and boulders where the water is shallow, and slow where the water is deep and tranquil. This endless variety is what gives a highland stream its attractive liveliness.

A change in depth automatically brings about a change in speed[23] because of the *principle of continuity,* which states that for any stretch of river that doesn't gain or lose water to the outside, the discharge across the upstream end of the stretch must be the same as the discharge across the downstream end. Or, what comes to the same thing, the volume that flows out in each second must equal the volume that flows in.[24] It follows that the flow must speed up wherever the cross-sectional area of the moving water body contracts, and slow down where it expands (see figure 5.8).

Whenever a river flows over a bed of boulders, so that shallow spots and

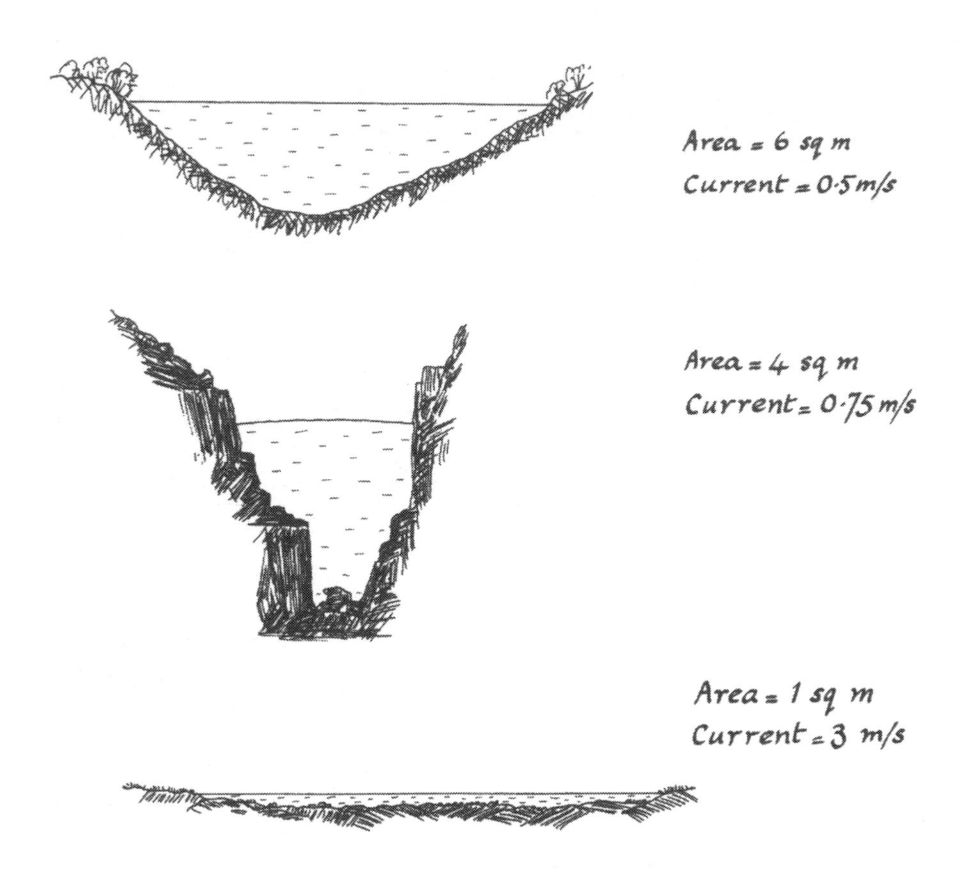

FIGURE 5.8. Cross-sections through an upland stream at three different sites, for which the cross-sectional area and current speed are shown. The discharge, 3 c.m.s. (cubic meters per second), is the same at each site.

deep spots alternate abruptly, the speed of each "thread" of water switches from fast to slow and back again every few centimeters; a wave crest forms wherever fast water is suddenly slowed, and a wave trough wherever slow water is suddenly speeded up. When all the threads making up a river form crests and troughs independently of one another, the result is *rapids*. The surface is steeply choppy, perhaps with some breaking waves. The choppiness is entirely unlike that raised by wind on stationary water—lakes and the sea.

The choppiness of rapids is unaffected by wind, and the individual waves, if they drift at all (some don't; see below), switch direction repeatedly. Here and there, a particularly large boulder on the riverbed will force the current to flow right up to the surface, forming an isolated patch of flat, swirling water amid the prevailing choppiness.

A sudden change in a river's depth does more than alter the speed of flow; it also alters the speed at which waves can move across the surface. In still water, waves travel more slowly where the water is shallow than where it is deep.[25] In moving water, too, the speed at which waves travel depends on the water's depth; and regardless of the depth, waves moving with the flow are speeded up and those moving against it are slowed down. Thus there are three different speeds to consider: first, that of the wave train as a whole, as it is carried along by the stream; second, the speed at which the waves would move if the water were still, which depends on the water's depth; and third, the speed of the waves *relative to the water,* which is the outcome of these two speeds combined and is known as the *wave celerity.*

The way flowing water behaves is visibly affected by whether the flow is faster or slower than the wave celerity. To see this, observe what happens if you dip a stick into a stream and hold it upright. If the speed of the current is less than the wave celerity, the streamflow is *subcritical,* or *tranquil* as it is also called. When this is so, a low, curved "collar" of little waves forms around the upstream side of the stick and remains there, wavering gently but not going anywhere; the wavelets spread left and right, forming a wide wake downstream of the stick. But if the speed of the current is greater than the wave celerity, the flow is *supercritical* (or *rapid* or *shooting*). In this case, the water piles up in a high, foaming collar before rushing past the stick to form a narrow, V-shaped wake (figure 5.9). Given a stream whose flow is subcritical near the bank and supercritical toward the center, it is possible to observe the change by testing with a dipstick at a sequence of distances from the bank.

At the changeover point—where the speed, as one might guess, is called *critical*—the sound of the flowing water changes, as well as its appearance: it becomes louder. This makes the changeover point easy to recognize when you are wading a fast stream: it is the point where the water suddenly starts to roar past your legs. Most streams flow at subcritical speed in most places; supercritical flow is the exception, normally occurring only when a stream is in flood, or in stretches where it is forced into a narrow channel.

Now consider what happens when the flow of a stream changes from rapid to tranquil or vice versa. These changes happen repeatedly where a fast stream

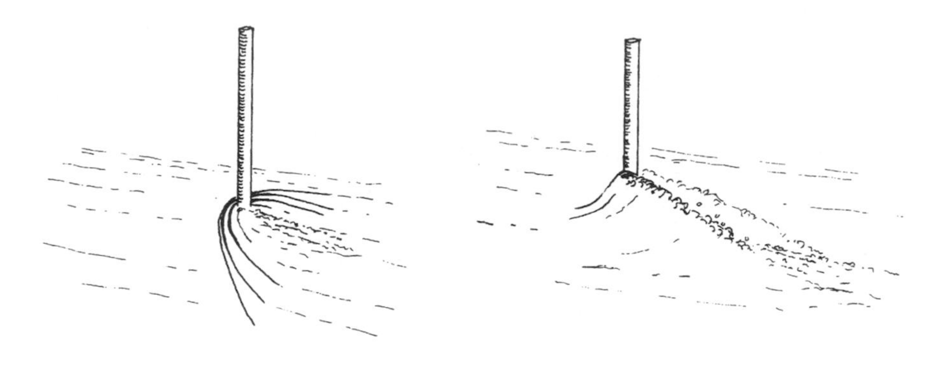

FIGURE 5.9. A stream flowing past a firmly held stick. On the left the flow is subcritical, as shown by the wide arc of tiny ripples curved round the stick's upstream side. On the right the flow is supercritical; the water froths where it touches the stick, and the wake is a narrow V.

flows down a channel strewn with boulders or interrupted by rock ledges. A change from supercritical to subcritical flow (a sudden slowing of the current) is a *hydraulic jump,* the reverse change (a sudden speeding up of the current) is a *hydraulic drop.*

A hydraulic drop is usually inconspicuous, but a hydraulic jump can be spectacular. When, at the foot of a waterfall or below a boulder, you see a standing wave that rears up as a foaming breaker with a tumbling, undulating crest that seems forever poised to rush upstream without actually doing so (figure 5.10), you are looking at a hydraulic jump. Waves form where the water suddenly slows down, but they cannot move upstream against the fast (supercritical) current speeding toward them; instead, they pile up until the crest topples, while the water foams and roars. The stronger the current upstream of the jump, the farther the water's momentum carries it before it is turned back on itself; in other words, the farther downstream of its cause the wave rears up. The current within the standing wave is directed upstream at the wave crest, making it a good takeoff point for salmon to leap from on their upstream migration to their spawning beds.[26]

A hydraulic jump is sometimes submerged below the surface and appears as a smooth mound with no breaking wave. The more deeply submerged the obstruction causing the jump, the lower the crest of the mound and the smaller the downstream gap between obstruction and crest. The way the hydraulic jumps change as the water level in a stream rises and falls is easy to

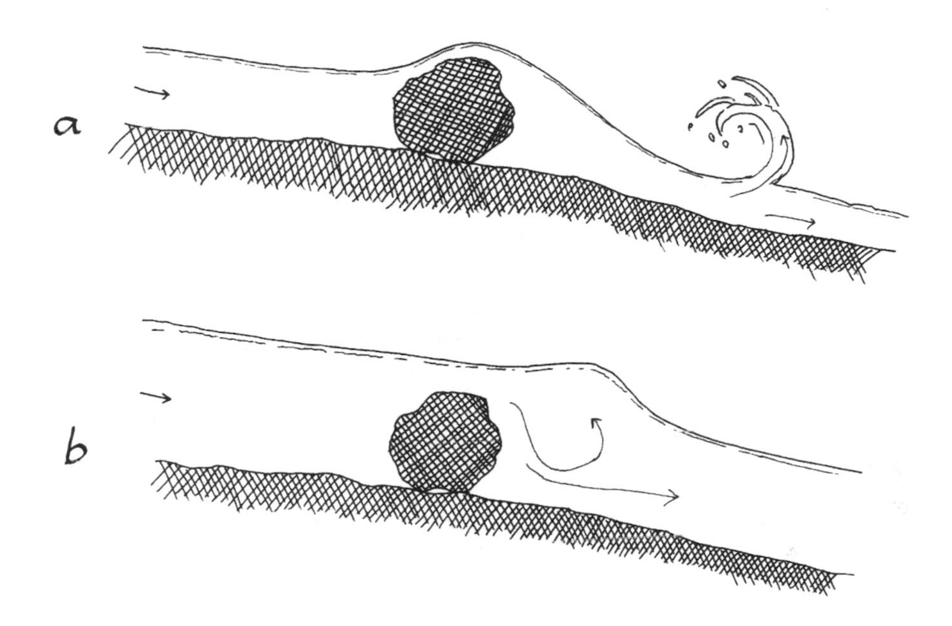

FIGURE 5.10. A hydraulic jump. *(a)* At low water, the jump is a breaking wave some distance downstream from the boulder causing it. *(b)* When the water level in the stream rises, the jump becomes a mound closer to the boulder.

observe in a stream you see frequently. For example, figure 5.10 shows the same stream when the level is low and when it is high. What had been a breaking wave in shallow water becomes a smooth mound when the water is deep, with the crest nearer to the obstruction.

Not all the waves in a stretch of rapids are hydraulic jumps. Some—those that wander back and forth instead of remaining fixed in one place as standing waves—mark changes in the speed of flow that do not bracket the critical speed, but remain below it or above it, usually below it.

The behavior of swift water when it meets a large obstruction such as a bridge pier suggests a quick, rough method of estimating a river's velocity. The bridge pier brings the water to a sudden, momentary halt, causing it to pile up before flowing away on either side, with a sharp downward dip at first (figure 5.11). The height from the top of the pileup to the bottom of the dip is labeled h in the figure. If you can measure h (in meters), you can

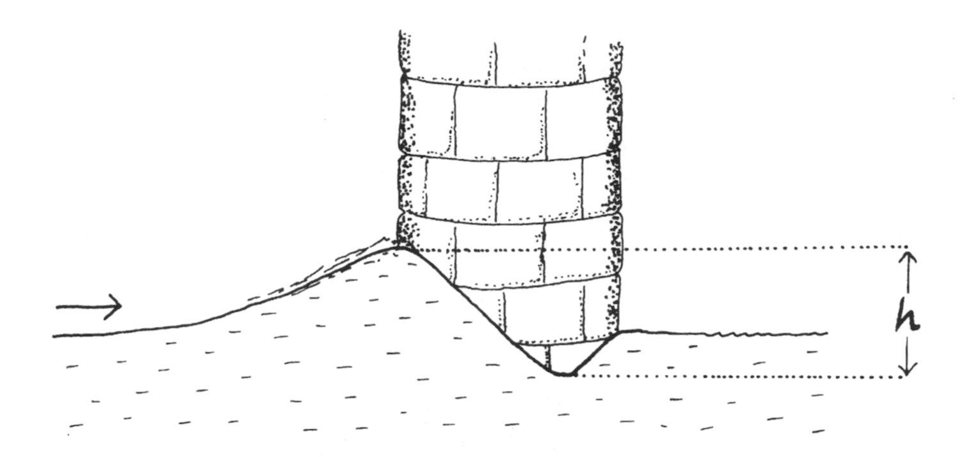

FIGURE 5.11. The disturbance to the flow of a river caused by a bridge pier. The drop from the highest water level (where the water first strikes the pier) to its lowest level as it flows past depends on the speed of the flow. If the height, labeled h, can be measured, the river's speed can be calculated.

calculate the water's speed V (in meters per second) from the equation $V = 4.43\sqrt{h}$. In real life this is not easy to do; the water level wobbles up and down continuously where it hits the bridge pier, making exact measurement difficult, even where it can be done at all; often the site is inaccessible.

To end this section on a technical note, I introduce the *Froude number* as a way of describing the speed of flowing water at a particular site. It is the speed of flow divided by the celerity of the waves at the site, which depends on the water's depth. The Froude number is less than one when the flow is subcritical, greater than one when it is supercritical, and (to state the obvious) equal to one when it is precisely critical.

5.9 A Closer Look at Flowing Water

Besides distinguishing between supercritical and subcritical streamflow, there is another distinction to consider, that between *laminar flow* and *turbulent flow.*

The difference is easy to see if you allow water to flow in a thin stream

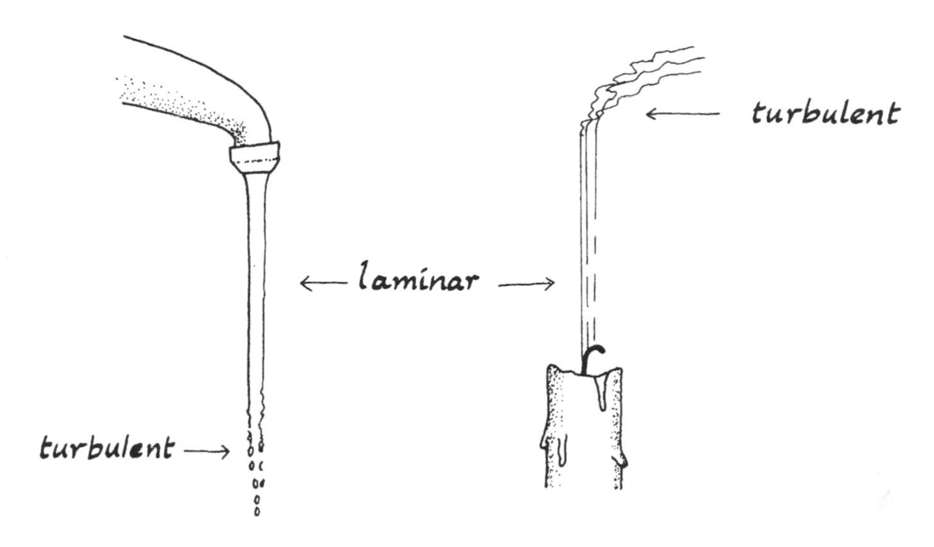

FIGURE 5.12. Examples of laminar and turbulent flow.

from a kitchen faucet. The upper part of the water column issuing from the faucet is smooth and regular with parallel sides; its flow is laminar. The lower part of the column is irregular and lumpy; its flow is turbulent. Moving air, too, can flow in laminar or turbulent fashion; everybody is familiar with the way a column of smoke rises from a freshly snuffed candle, with straight, vertical threads rising steadily for some distance and then, suddenly, becoming haphazard and tangled (figure 5.12).

It is easy to describe what is happening in general terms. In laminar flow all the molecules are moving in parallel straight lines. In turbulent flow numerous "threads" of water (or air, as the case may be) are following their own paths: swirling eddies form and dissipate, rolling over one another and becoming tangled and mixed. Laminar flow can be described scientifically, in terms of mathematical formulas, but there is no precise definition of turbulent flow.[27] All that can be said is that it is a splendid example of chaos, in both the colloquial and scientific senses of the word. Scientifically speaking, chaos is a state of perpetual instability, in which every moving particle follows its own unpredictable course. A good way to appreciate the chaotic behavior of turbulent water is to watch the behavior of the standing breaker at a hydraulic jump; the water rears, wobbles, bubbles, and tumbles, in an unforeseeable way. You can watch for hours; the wave as a whole doesn't shift, but its internal structure goes through endless changes of pattern that seem never to repeat themselves.

Indeed, nearly all the flow in streams and rivers is turbulent. It is laminar only in a thin layer over very smooth surfaces, in the narrow interstices between rocks and boulders, and among the stems and leaves of water plants.

The contrast between laminar and turbulent flow has nothing to do with the contrast between subcritical and supercritical flow described above. Flow can be laminar at both subcritical and supercritical velocities, and likewise for turbulent flow.[28]

Now consider what it is that determines whether the flow of a liquid will be laminar or turbulent. Everyday observation shows that laminar flow is seen much more often in syrupy or oily liquids than in water, and this suggests one of the controlling factors: the *viscosity* of the liquid.

All liquids—including water—are viscous to some extent. Viscosity is the tendency of a liquid to "stick to itself" and, therefore, to resist a shearing force, in other words, to resist the sliding of one layer over another; but liquids also stick to solids, even more strongly than they stick to themselves. This leads to an astonishing conclusion, namely, that water can*not* slip over a solid; it clings to it. Consequently, right at the bed of a stream, the flowing water isn't moving at all. All movement is entirely within the water and there is no slippage whatever between the water molecules and the solid material over which it is flowing. This astonishing state of affairs is called the *no-slip* condition; it holds for all solid surfaces, no matter how slippery they are. The no-slip condition runs counter to common sense and was the subject of fierce disagreement in the nineteenth century, but its truth is supported by the most precise observations currently possible.[29]

An inevitable consequence of the no-slip condition is that there can be no friction between a stream and its bed and banks. Bed and banks do, undoubtedly, resist the flow of water, but the resistance is not friction; rather, it is resistance to shearing between adjacent layers of water and takes place entirely within the water.

The way water flows is this: a microscopically thin layer of water clings immovably to all submerged solid surfaces. Imagine that the only solid surface is a smooth, flat streambed, and that a layer of motionless water is clinging to it; then layers at progressively greater heights above the streambed slide over the layers below them at progressively greater speeds, each dragged by the layer above it; the dragging results from the tendency, already mentioned, for water to stick to itself. A short distance (sometimes microscopically short) above the bottom, this layered, laminar flow gives way to chaotic, turbulent flow. In a stream with an unusually smooth bed—for instance, of polished bedrock, or of exceedingly fine-grained mud—the whole bed may be overlain by an extensive *laminar* or *viscous layer*[30] whose thickness depends on the velocity and temperature of the water. The faster the flow and the warmer the water, the thinner the laminar layer will be. In water at 15°C, for example, the laminar layer is only about 5 millimeters thick even if the water flows so slowly that it takes 20 seconds to cover 1 meter; if the flow speeds up to 1 meter per second, the layer will shrink to a mere 0.25 millimeters.[31] At a given flow rate, the layer is thicker in cold water than in warm because the viscosity of a liquid depends on its temperature, being greater at low temperatures than at high (a fact understood by anybody with the foresight to chill whipping cream before attempting to whip it).

A widespread laminar layer can form only over a streambed so smooth that all surface irregularities are small enough to be submerged in the layer. The bed is then described as *hydraulically smooth*. Streams and rivers rarely have such beds. Much more often, the beds are *hydraulically rough*, with bigger irregularities. The irregularities break the continuity of the laminar layer, and turbulent eddies form among them. All the same, a microscopically thin layer of water in contact with any solid surface must flow in laminar fashion because of the no-slip condition. A river with a hydraulically rough bed that a hydraulic engineer might regard as "turbulent right down to the bottom" would appear quite different to a biologist studying aquatic microorganisms so small that they lived permanently submerged in the microscopically thin laminar layer coating each tiny irregularity.

In some circumstances laminar flow occurs at the surface of a stream, as well as at the bottom. This happens when an obstacle—a boulder, or a stick, for instance—protrudes through the water surface. The obstacle must be blunt so that it impedes the water, rather than knife-edged so that it slices through. Then the water is brought to a momentary standstill where it hits the obstacle and, as well, in a patch of "dead water" immediately downstream of it; after the halt the water gathers speed again. The result is a complicated, small-scale pattern of slow, laminar flow and fast, turbulent flow around the obstacle, at the surface.

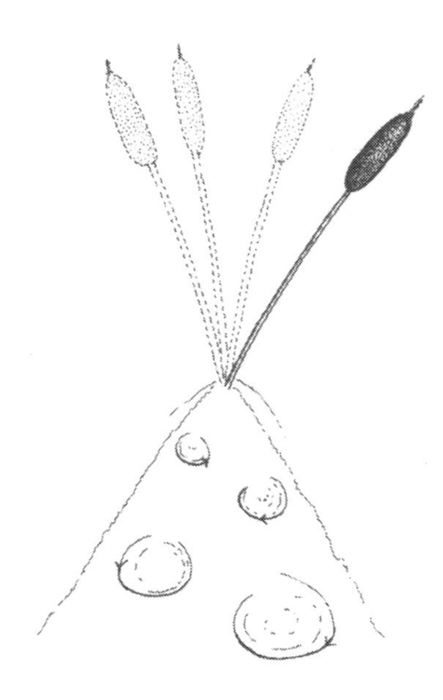

FIGURE 5.13. The paired rows of vortices formed in the wake of an obstacle, in this case a cattail, when water flows past it (the flow is toward the viewer). The cattail is shaking vigorously, as it is pushed alternately to left and right by the vortices.

If the obstacle is a vertical cylinder, such as a willow branch or a cattail stem in a flooded watercourse, the result is *unsteady laminar flow,* in which a series of vortices form (see figure 5.13). As they flow past the cylinder, the vortices are "shed" alternately to left and right, leaving a train of them downstream of the cylinder, with those on one side rotating in the opposite direction to those on the other.[32] Each vortex pushes the cylinder aside as it goes by. This is what is happening when you see a stick protruding through the surface of a fast stream wagging rapidly from side to side in spite of the fact that the "push" of the water seems to be directed entirely in the downstream direction. If the vortices push the stick alternately left and right at a rate coinciding with the stick's natural vibration rate, a conspicuous, regular wagging motion is set up, which has mystified many a naturalist who chanced to notice the effect.

6

Rivers at Work

6.1 Rivers as Conveyor Belts

Rivers and streams are tremendously varied: some are swift, others sluggish; some are clear, others muddy; some have carved channels in hard rock, others in soft sediments. Indeed, no two rivers are exactly alike.

The way a river looks and behaves depends on the terrain through which it flows—whether the slopes are steep or gentle, whether its channel is in solid rock or in loose, unconsolidated material—and on the *sediment load* it carries. This load is the mineral material carried along by the flowing water; it consists of everything from boulders to rock dust to tiny clay particles formed by the chemical decay of rocks, plus small amounts of organic material such as dead leaves and twigs; it is the material that becomes *sediment* or *alluvium* once it has come to rest. (The word *sediment* is sometimes used to mean all deposited material, regardless of whether it was transported by water, wind, or glacial ice. *Alluvium* means sediments deposited by rivers.)

All rivers and streams, even the clearest, carry a load of sediment at least some of the time. Rivers are conveyor belts: they are forever picking up solid matter in one place and putting it down in another. They are not merely bodies of flowing water, but of flowing water-plus-sediment. A river sometimes erodes the land, thereby increasing its sediment load, and sometimes builds up new land by depositing some of the material it is carrying. These processes result in degradation and aggradation. *Degradation* is the wearing away of the land by any of a number of agents: rivers, glaciers, ocean waves, and the wind, of which rivers are by far the most important. It has been estimated that, on average, the ground level of the whole world is worn down at the rate of 8 millimeters per century.[1]

Aggradation is the opposite of degradation: it is the creation of new land, or the building up of existing land, by the deposition of some of the material picked up and carried by the agents of degradation: rivers create floodplains and deltas, winds create dunes. Aggradation was more productive during and immediately after the last ice age; vast quantities of material froze to the bottom of creeping ice sheets and were carried with them and laid down elsewhere as *glacial till.* Once the ice had melted but before the land became vegetated, the wind picked up, and subsequently put down, vast quantities of the loose dust, consisting of finely ground rock flour, left exposed when the ice disappeared. This wind-borne dust created enormous beds of the soil known as *loess.* Nowadays, however, ice and wind are of negligible importance compared with flowing water as agents of aggradation. Moreover, aggradation does not balance degradation so far as the world's land surfaces are concerned; eroded surface material carried out to sea in rivers, or in the backwash of ocean waves, is lost to the land, if not forever, then until crustal warping or a fall in sea level brings it up into the open air again.

6.2 The Work of Rivers: Erosion

The sediment load in rivers and streams comes from two sources. Some of it—chiefly mud—dribbles in with the overland flow, the water that drains directly off the land into the nearest river; and some comes from within the river's own channel. It is washed out of the channel walls by the river itself. The material of which the channel walls are made determines what the river will be like: whether it will be a crystal-clear stream in a channel carved through hard rock, or a muddy river in a channel carved through alluvium. An alluvial river flows in a channel cut in material laid down by an earlier river, most likely by itself. As we shall see (section 6.5), rivers in flat country continually shift sideways, from left to right and back again: because of this, a river often shifts back over old ground, land it created itself, decades or centuries earlier.

Alluvium erodes much more easily than hard rock, but even the clearest mountain streams erode their channels gradually; the smoothly polished rocks that often form part of their beds prove that all rocks, however hard, can be eroded. The question is, how does it happen?

The first point to notice is that pure water cannot wear away solid rock. Rock can only be worn away by abrasive sediments carried in moving fluids.[2] Pure water cannot wear away anything, as it does not *slip* over a solid surface (see section 5.9); rather, a microscopically thin layer of the water clings to the solid, and all the slippage takes place within the water itself. This rule applies to all fluids, gases as well as liquids; it explains why you cannot blow all the dust off a smooth surface but must wipe it off with a rag. Water can erode solid rock only if it strikes it with "missiles," sand, grit, pebbles, and larger rock fragments that it picks up and carries.

This raises the question, how heavy a rock fragment can a given stream transport? To say the same thing in technical terms: what is the stream's *competence*? The answer obviously depends on the speed of flow: the faster the stream, the bigger the rocks it can roll along the bed or carry above it.

Consider rocky streams first, in which the fragments to be moved are big and easily seen in the clear water. In round figures a current of 15 centimeters per second is enough to shift coarse sand; 50 centimeters per second shifts medium gravel; and current speeds of, respectively, 2 meters per second (m/s), 4 to 5 m/s, and 6 to 7 m/s will shift medium cobbles, medium boulders, and large boulders.[3] These are the speeds at which the water must flow close to the streambed, where the rocks to be moved are lying.

Obviously, it would take a large, fast river funneled by a rocky gorge to shift anything as large as a large boulder. All the same, large boulders are often seen lying on the beds of quite modest mountain streams. How did they get there? The answer is that, when a stream cannot move a boulder, it can often undermine it; the flowing water washes away the finer gravel and sand supporting it, whereupon gravity causes the boulder to roll down to the bottom of the channel. In this way heavy boulders and cobbles come to rest on a smooth, polished streambed of solid rock, a scene familiar to all mountain travelers. The polishing is done by rock fragments small enough for the water to carry or roll. They strike the larger "immovable" rocks, and every collision does a tiny amount of damage to both of the colliding rocks, the moving one (the "missile") as well as the stationary one (the "target"). Both become scratched, ground, and polished; the process is known as *abrasion.*

Erosion in an alluvial channel proceeds differently. The sediments forming the banks are fairly soft. When the water level is high and the banks are submerged, they become saturated with water; when the water level falls, the pressure of this "trapped" water causes chunks of the bank to break away and fall into the river. The chunks disintegrate in the flowing water and are carried off as dispersed mud and sand. In many alluvial rivers this is the chief cause of erosion, but it is not always the only one; as well, the water will pick up additional mud and sand from the bed and banks unless it is already overloaded (see section 6.3). In any case, *erosion,* defined as the wearing away of the land by mechanical forces, can take place in more than one way; how it happens depends on circumstances.

The picking up of fine sediments (fine sand, silt, and clay) by flowing water does not, surprisingly, become easier as the particles become finer. The material most easily *entrained* (picked up) is medium sand, with grains about 0.5 millimeters in diameter. Finer sediments require a faster current to lift them off the bottom, even though they are lighter in weight. This is because the smaller the particle, the more likely it is to be within the layer of laminar flow close to the bottom and hence out of reach of the turbulent eddies that make it easier for moving water to pick particles up. In addition, the exceedingly fine particles of silts and clays are attracted to each other by electrical forces that make them stick together.

The way in which a river carries its load depends on the speed of the river's flow, and on the mix of coarse and fine particles. The heavier particles, which roll and bounce along the bottom, are known as *bedload;* the finer particles, which are held in suspension in the water and flow along with it, constitute *suspended load.*[4] As a river level changes with the seasons, the proportions of these two kinds of load changes: a particle that behaves as bedload at low water may be lifted into the suspended load when the river is in spate. The relative amounts of the two kinds of load also change as a river flows from its source to its mouth: near its source in the uplands, where the river is clear, it contains far more bedload than suspended load; the opposite is true when the river reaches the lowlands and becomes muddy.

6.3 *The Work of Rivers: Sediment Transportation*

A river's load of sediment is carried only as long as the river is not overloaded. Picking up a load of sediment and carrying it along is work, in the physicist's sense of the word *work* as well as colloquially. The rate at which a river performs its work, or the rate at which it uses up energy (the two phrases mean exactly the same thing), is its *power.* Imagine a stretch of river along which no tributaries enter, so that, from the principle of continuity, the discharge is the same at the downstream as at the upstream end; the river's power in the stretch depends on two things: the discharge (the volume of water entering and leaving the stretch per second) and the height it loses going from the upstream to the downstream end. This power controls the amount of sediment load the stretch of river can carry, an amount known as its *capacity*. Capacity doesn't depend only on power, however: for given gradient and discharge, the speed of flow close to the bed, where the sediments lie, is greater in a wide, shallow stream than in a narrow, deep one; this means that (provided all other factors are equal) the shallow stream will have greater capacity.[5]

Rivers are not always full to capacity with sediment. The load a river actually carries depends on how much "pickupable" material is available to it. Its capacity merely sets an upper limit. In the same way that a river has a discharge rate, which is the rate at which its water flows past a given point on the bank, so too it has a *sediment discharge rate*, the rate at which sediment flows past a point. The sediment discharge rate depends on the ordinary (water) discharge rate and on the landscape the river flows through; an alluvial river flowing over a muddy floodplain obviously picks up more sediment—it may be full to capacity—than a mountain river. Consequently, the sediment concentration, usually measured in parts per million (ppm),[6] is much greater in alluvial than in bedrock rivers. This is obvious in a general way to anybody; here are a couple of examples of the actual quantities involved.

Where the Snake and Clearwater Rivers emerge from their mountain valleys and unite near Lewiston, Idaho, the sediment concentration is about 10 ppm at low flow (500 c.m.s.) and about 25 ppm when the discharge is twice as great (1,000 c.m.s.). The corresponding numbers for the Tanana River, Alaska, measured near Fairbanks where the river is flowing over a wide floodplain, are about 500 ppm and 1,500 ppm.[7] As you might expect, an increase in discharge rate causes a more marked increase in sediment in the muddy river than in the clean one; the Clearwater in Idaho was not so named without reason. Very muddy rivers in spate can have a sediment concentration as high as 20,000 ppm, or 20 kilograms per ton.[8]

These numbers refer to sediments passing a given point on their respective rivers. Geography books often give the sediment yield for whole watersheds, which is the sediment discharge at the mouth of the watershed, usually where it meets the sea. The numbers are millions of tons per year. Some examples:[9] for the Yangtze, 500; the Amazon, 498; the Mississippi, 312; the St. Lawrence, 4. The *Hydrological Atlas of Canada* gives the sediment yield of the

Fraser River of British Columbia, the so-called muddy Fraser, as a volume: if a year's supply of sediment at the river's mouth were shaped into a cubical block, its height and sides would be about 300 meters, a truly gigantic size, weighing over 70 million tons.[10]

6.4 When Work Stops: Deposition

As we saw above, a river's capacity to shift sediment is limited. The capacity varies along the river's course; at every point, capacity depends on the speed of flow, and where a river loaded to capacity slows down, because it meets a barrier or because the land slopes more gently, some of the load is dumped. In a word, *deposition* takes place, beginning with the heaviest particles in the load; in this way the load decreases until it no longer exceeds the river's capacity to carry it.

As to the individual particles, deposition proceeds from heaviest to lightest. Deposition differs from erosion in this respect: recall that flowing water picks up fine sand more easily than silt, and silt more easily than clay notwithstanding the decreasing weight of the particles. When sediment is deposited, the heaviest particles become grounded first and the lightest last; this is because the lighter the particle the longer it takes to settle out, and the farther it will drift in the current before it eventually reaches the streambed. The finest sediments are not deposited at all unless the flow comes to a complete stop, and they may take months to settle out when that happens.

Deposited river sediments create "structures" of tremendous variety, ranging in size from the tiniest ripple of sand on the riverbed to floodplains and deltas measuring tens of thousands of square meters in extent.

Let's consider small-scale effects first, and consider what happens when sediments are picked up and carried for only a few meters, or even a fraction of a meter, before being deposited again (figure 6.1).[11] Given a smooth bed of sand, a gentle current will form it into *ripples,* small ridges lying across the current at right angles. Ripples are usually lower than 4 centimeters in height and less than 60 centimeters from crest to crest: the finer the sand the sharper the crests. The sand picked up by the current as it flows up the gently sloping upstream side of a ripple is dropped on the steeper, lee side where the speed of flow decreases suddenly. This makes streambed ripples asymmetrical; they differ in this respect from the symmetrical ripples formed on a sandy beach by the back-and-forth flow of waves. Riverbed ripples migrate downstream, as the current carries the sand.

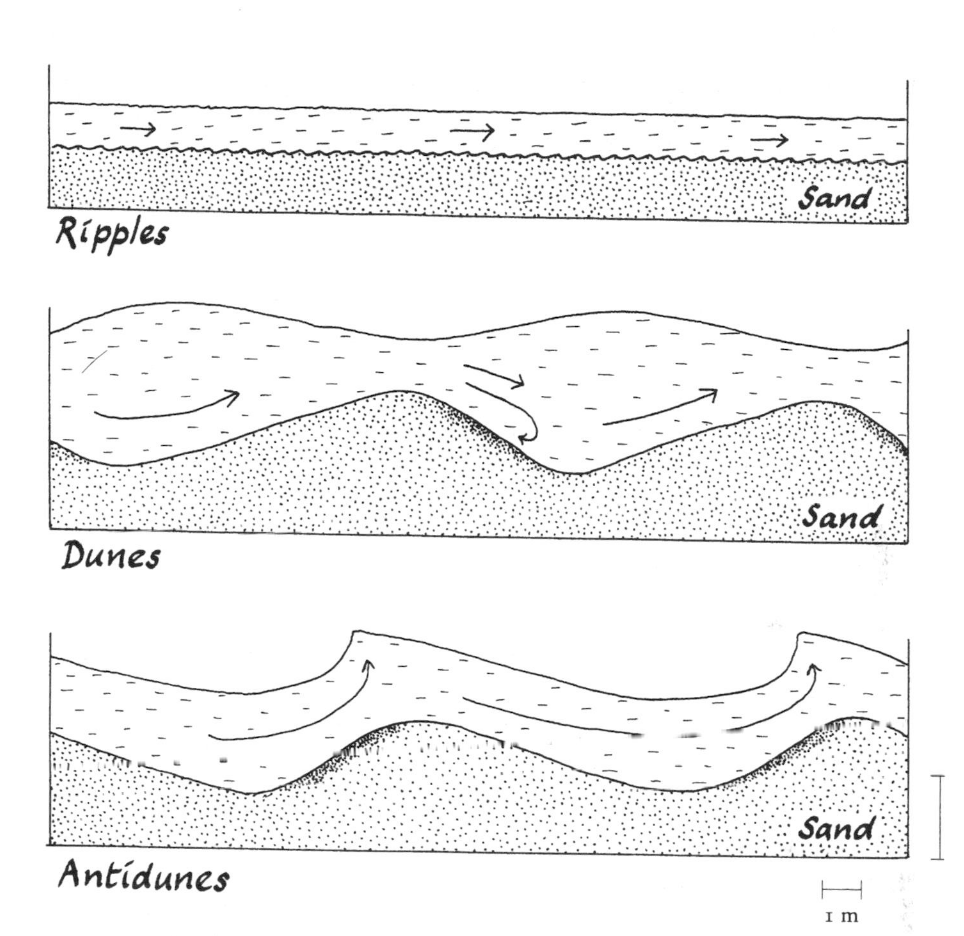

FIGURE 6.1. Ripples, dunes, and antidunes. The drawings are all on the same scale, which is shown at bottom right; note the 1-meter scale bars, showing that the vertical scale is twice the horizontal scale. Where dunes develop, the surface of the water is slightly lower above the dune crests than above the troughs. The reverse is true with antidunes: the wave crests in the faster-flowing water are above the antidune crests. Heavier stippling shows where sand is being deposited.

Deeper, swifter streams flowing over rougher, but still sandy, beds create *dunes* (piles of sand heaped up by moving water are just as much dunes as are piles of dry sand heaped up by the wind). Underwater dunes look like outsize ripples, but the physics of their formation is different.[12] Depending on the depth of the water, they may be more than a meter high and have a wavelength (the distance from one crest to the next) of over 10 meters. Like ripples, they migrate downstream. They are built by turbulent eddies, and their crests are rounded; if the flow speeds up, the dune crests sometimes become flatter and lower until they are flattened out of existence. But sometimes, if conditions are right, *antidunes* are created.[13] This happens if, in a heavily loaded, fast-flowing river, transverse, breaking waves form on the surface with their crests directly above the crests of the dunes on the bottom. This arrangement is precarious, and seldom lasts long; but while it does last, turbulent eddy currents erode sediment from the downstream side of each erstwhile dune and deposit it on the upstream side, causing the upstream slope to become steeper than the downstream slope; the resultant antidunes migrate upstream. So antidunes are "anti" to ordinary dunes both in shape and in their direction of migration (figure 6.1).

Dunes and antidunes are usually submerged in turbid, sediment-laden water. Larger, more visible sediment deposits are *bars,* low ridges that show above water at least some of the time. In slow rivers the bars are usually aligned across the current; these are *transverse bars.* A possible explanation for them[14] is that speed and turbulence fluctuate slightly along any given stretch of river and the material that builds transverse bars drops out of the load wherever the speed is at a low point in these imperceptible fluctuations. In a really big river the bars are big, and the "wavelength" of the fluctuations long; in the lower Mississippi the bars are 8 meters high and the wavelength (the distance from bar to bar) is 230 meters. Material eroded from the more gently sloping upstream side of each bar is deposited on the steep lee (downstream) side, causing the bars to migrate gradually downstream. In a sense, they are oversized dunes.

In swift rivers long, diamond-shaped, *midchannel bars,* aligned parallel with the current, are found. They are particularly common in mountainous country and in the Arctic, where rivers flowing from glaciers are in spate following spring breakup into midsummer. The rivers are powerful, and their erosive power is tremendous: added to the ordinary processes of erosion is *thermoerosion,* the simultaneous melting of frozen riverbank mud and its washing away that comes with spring breakup.

Huge quantities of rock debris, ranging from boulders to clay, deposited by glaciers, are available and waiting to be transported. The result is often a swarm of midchannel bars so big that it fills most of the riverbed, with open water occupying only a fraction of it (figure 6.2a). A river in this state is called *braided.* Braided rivers are wide and shallow, and the bars tend to be unstable: they are repeatedly washed away and then built again in different places, especially those in midstream; those nearest the banks sometimes last long enough

FIGURE 6.2. River bars. *(a)* A braided river with numerous unstable, shifting bars of boulders, cobbles, gravel, and sand (Weasel River, Baffin Island). *(b)* A deeper river, less prone to flooding; a large bar has become stabilized by growing plants to form a permanent wooded island (Fishing Branch River, Yukon Territory).

to allow a few plants to take root on them, but those near the middle are no more than low mounds of sand, gravel, and cobbles, wholly devoid of life. A map showing details of the individual bars would not be the same from one year to the next.

Elongate midchannel bars also form, in smaller numbers, in less dramatic rivers, that is, quieter, narrower, deeper ones, in milder climates. A bar in such a setting is usually composed of sand, and is much more stable; instead of being periodically washed away, it may become a permanent island on which vegetation develops (figure 6.2b). The plants contribute to the stability: their roots bind the sand, and the withered remains of dead leaves continually thicken the developing soil by adding a fresh layer of organic material every year.

Returning to a consideration of braided rivers, they are the place to look for good examples of *sorted sediments,* those in which particles of different sizes have become sorted or separated because of their different weights. In midchannel bars the different sediments are deposited one after another: bars of cobbles and gravel often have "tails" of finer material, sand and silt, at their downstream ends. The current has been so slowed by its passage over the upstream, gravel end of each bar that it can no longer transport the lightweight sand and silt, which are deposited at the downstream end. Sediments are often sorted into horizontal layers, too. A stream in spate, during the rainy season or while the snow thaws, will deposit only the coarse fraction of its load and carry away the fine fraction; not until the current subsequently slows down again will the fine sediment settle out. But the settling *fines* (sand and silt) must first fill the interstices among the individual cobbles and pebbles of the coarse layer before they can form a blanketing layer on top.

Observing how a river varies from season to season reveals how its sediment load constantly changes, in composition as well as in amount. Particles of a great range of sizes are picked up, carried, and put down. As we noted at the beginning of this chapter, a river acts as a conveyor belt. But it differs from a conveyor belt in one important respect: the load moves slower than the belt.

The load is not a single entity: it is continually augmented by material newly eroded from the river's bed and banks, and continually depleted as its heavier ingredients are deposited. At the same time the entrained material is being continuously sorted: the fine particles flow as fast as the water itself, while the coarser particles roll and jump along the bottom, and then remain still for a while. Whereas every particle moves individually, at its own speed, the load as a whole can be said to have an average speed. Hydrologists have estimated that in a typical alluvial river, where the water flows at 1 meter per second (3.6 kilometers per hour), the train of alluvium flows at a mere 1 meter per hour, on average.[15]

Up to this point we have considered how deposited sediments create small and medium-sized structures (such as ripples, dunes, and bars) *within* a river's channel. A large river, loaded to capacity with sediment, can create structures extensive enough to dwarf the river that produced them. Then the river channel is within the sediment instead of the other way around; the rivers flowing across floodplains, or the splayed channels crossing a delta are familiar examples. We return to these large-scale structures in section 6.9. Before that we take a look at the medium-scale consequences of erosion and deposition happening together.

6.5 Winding Rivers: Meanders

A straight river is a most uncommon sight; rivers nearly always have winding courses. The windings of alluvial rivers often take the form of *meanders;* that is, rather than being random and unpredictable, the windings are fairly regular, with rounded curves looping to left and right alternately (figure 6.3).

Anybody who enjoys studying maps (or merely gazing at them) is familiar with the meander patterns of rivers flowing across flat lowlands, and must

FIGURE 6.3. A meandering river.

have noticed that they often have much the same shape regardless of the size of the river: the pattern is on a scale that matches the river's size (figure 6.4).[16] The wavelength of the meanders seems to be controlled by the width of the river's channel; it is nearly always between ten and fourteen times the channel width,[17] averaging about eleven channel-widths. (The wavelength is the straight-line distance across one complete **S** of the winding course.)

What causes meanders to develop? This has been fiercely debated by hydrologists for many years. The commonsense explanation is as follows.

First, visualize what happens if you strike a tall but not very strong garden plant with the jet from a garden hose (most gardeners have done this inadver-

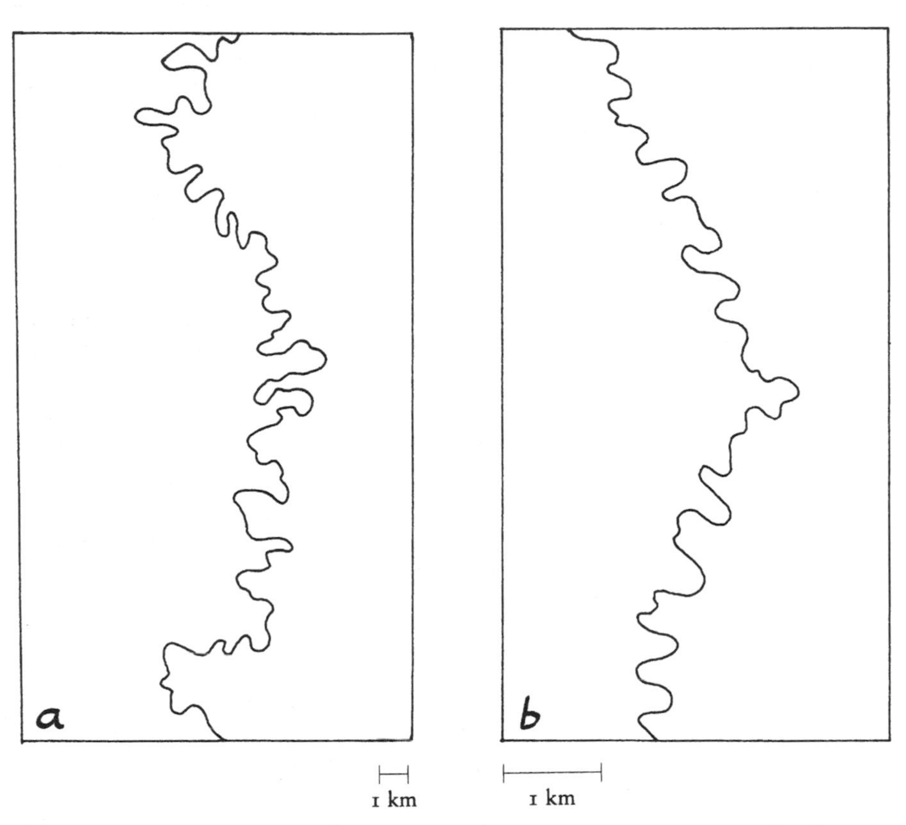

FIGURE 6.4. Two very similar meander patterns: *(a)* the Old Crow River and *(b)* its tributary, Thomas Creek, in Yukon Territory. Notice that, in spite of their overall similarity, the scales of the two patterns are very different, as shown by the 1-kilometer scale bar below each map.

tently many times!): the jet of water is deflected, and at the same time the plant bends back from the blow. That is, both the striker (the water jet) and the struck (the plant) have "given" to some extent: the energy in the original jet has been split to yield two outcomes. If the energy of the jet and the strength of the plant were accurately known, it would, in theory, be possible to predict how much the plant would bend and through what angle the jet would be deflected.

Now imagine the situation in an alluvial river. The current is a jet of water that here and there strikes the bank on one side or the other (the banks are never geometrically straight in a natural river). Suppose it strikes the left bank, that is, the bank on the left as you face downstream. The current is deflected to the right, and simultaneously undercuts the bank, dislodging some of the bank's unconsolidated gravel, sand, and silt. The deflected current strikes the adjacent section of bank more obliquely, undercutting it a little more and at the same time being deflected a little farther to the right. The process continues, until eventually the current has been deflected so far that it impinges on the opposite (right) bank of the river; next, undercutting of the right bank, accompanied by deflection of the current to the left, takes over. And so on and so on. This is the way in which a meandering river acquires its alternating loops.

The commonsense account does not explain why meanders have the shape they have, nor why the meander wavelength should be about eleven times the width of the river. To explain these things, it would be necessary to know how the energy in river water that strikes the bank a glancing blow is partitioned: some of the energy is "used" in undercutting the bank, redirecting the current and transporting sediment; some is "wasted," that is, converted into heat, which warms the water very slightly; and some is "saved" and carried away in the flowing water; it would also be necessary to allow for the way the partitioning continuously changes, as the water rounds each curve.

These problems have given scope for much scientific enquiry into the details of the meander-forming process. The underlying scientific law is presumably the *principle of least action.*[18] Keep in mind that a scientific law is not a law that nature obeys: nature doesn't "obey" anything. Rather, a scientific law is a description of an underlying mechanisms believed, in the light of current knowledge, to explain events in the real, physical world. The principle of least action explains a tremendous number of natural events; Newton's laws of motion and the laws describing the behavior of electromagnetic waves are all spin-offs from the principle.

Comparing the principle's predictions with real life is difficult, however, because the meanders of a natural river are never geometrically perfect. Irregularities are inevitable, for several reasons. The river's discharge rate—hence the force with which it strikes its banks—varies with the seasons. The strength of its banks (their resistance to undercutting) varies from place to place, and changes from time to time: riverbanks weaken when sodden by rain, and harden during a dry period. As a general rule, a river with weak banks tends to become wide and shallow, and its meanders tend to be long; conversely, if its banks are strong, a river is likely to be relatively narrow and deep, with short meanders.

With conditions as variable as they are in nature, it is remarkable that river meanders are as regular as they often are. Moreover, they remain regular even while they are forever changing: meanders never stay the same for long because the undercutting that produces them never stops.

Sometimes, although the overall shape of the meander pattern stays roughly the same, it shifts downstream bodily. This happens when undercutting of the banks is strongest just around the corner of each curve (see figure 6.5), and some of the sediment load (picked up when the river undercut the bank at the previous bend) is likewise deposited just around the corner of each lobe of land. The meander pattern, as a whole, migrates downstream.

A river behaves differently when the undercutting is strongest at the center of each concave curve and the sediment is mostly deposited at the tip of each protruding lobe (figure 6.6). When this happens, the meander pattern tends to become more and more exaggerated. It may become so rococo that two loops of the channel expand to meet each other, or equivalently, that the neck of the lobe between them narrows to nothing, whereupon the river takes a shortcut. The shortcut is a *neck cutoff,* and the arc of still water now separated from the flowing river is an *oxbow lake.* This is not the only way a meandering river can shorten its course: at times of flood, or if the banks are weakened, for example by clearcut logging, a river may cut a new channel through the neck of a meander lobe while the neck is still wide; the result is a *chute cutoff.*

Perhaps the most regular of all meander patterns are those made by meltwater streams flowing off a glacier (figure 6.7). They owe their perfection to the uniformity of the ice: ice is obviously much more homogeneous than alluvium.[19]

A notable characteristic of glacial streams is the purity of their water. Glacial meanders demonstrate that abrasion by water-borne sediments is not necessary for meander formation. On glaciers there is no undercutting of the

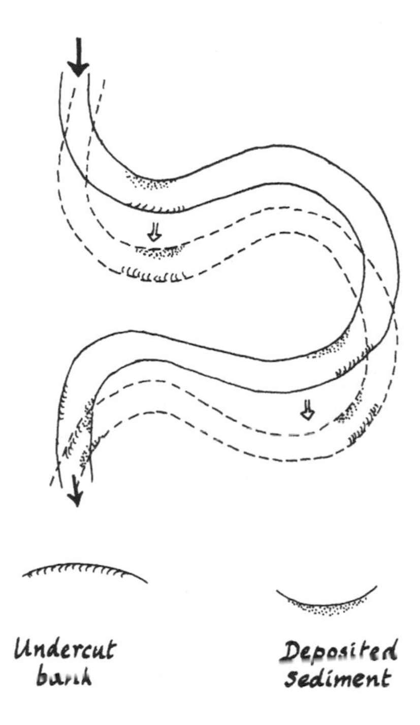

FIGURE 6.5. A migrating river meander. Solid lines show the banks as they were; dashed lines show their new locations. Note that the undercutting and the sediment deposition are both most pronounced a short way downstream of the axis of each bend; the meander migrates downstream.

banks either, that is, no dislodgement or breaking off of projections, for in a glacial stream there are no projections: the whole channel is always dangerously slippery. The wearing away of a stream's icy banks that is a necessary part of meander development can only result from thermoerosion, the melting of the ice by flowing water. The heat comes from the "wasted" energy mentioned above, generated whenever flowing water is deflected.

Returning to ordinary alluvial rivers, it is worth noting that when meanders expand or shift they sometimes carry political boundaries and property lines with them, much to the dismay of people unable to comprehend and foresee the power and downright lawlessness (in the human sense) of nature.

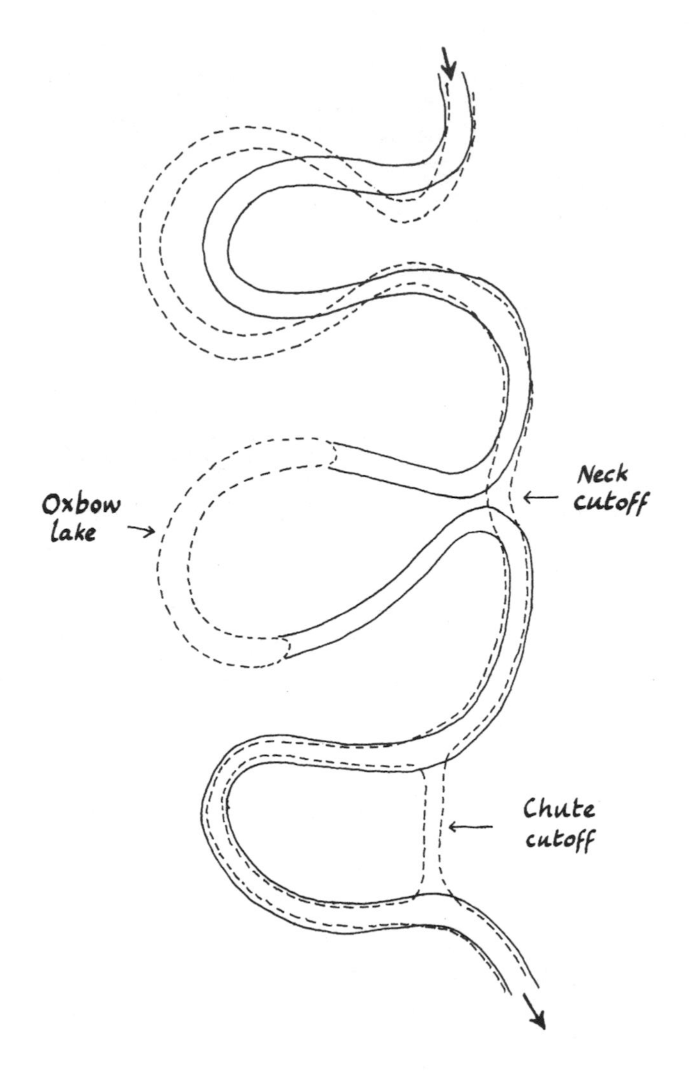

FIGURE 6.6. Three developments of a meander pattern; earlier and later states shown by solid and dashed lines, respectively. Upstream (at top) a meander tongue is lengthening while its neck narrows. At the center a neck cutoff has cut through the base of the tongue, leaving a stagnant oxbow lake. Downstream (at the bottom) a chute cutoff has created a new channel across the base of a meander tongue; when the discharge is high, water flows through both the cutoff and the original bend.

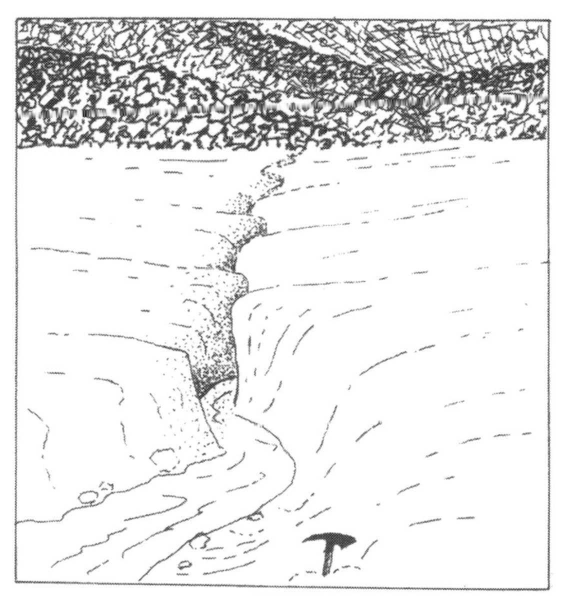

FIGURE 6.7. *(a)* A view of meltwater streams flowing off a glacier in the High Arctic (Ellesmere Island, Canada); the human figures show the scale. Note the extreme regularity of the meander patterns. *(b)* Close-up of a glacial stream; the ice axe in the foreground shows the scale. Note the smoothness and roundedness of the channel walls, incised in the ice by thermoerosion; the stream surface is about 2 meters below the general level of the glacier.

In the lower Mississippi the lobes of some of the meanders have lengthened by about 6 kilometers over the past 500 years.[20]

6.6 Meanders in the Vertical

Up to this point we have considered how meanders appear as seen from above. Figures 6.5 and 6.6 are maps; they reveal nothing about the ups and downs of the riverbed and how the depth of the water varies.

If you wade a meandering stream at a point where it curves most sharply, you will find that the water is deepest close to the undercut bank and becomes progressively shallower as you approach the *slipoff* bank directly opposite; the slipoff bank, around the outer end of a meander lobe, is low and gently sloping (figure 6.8). The current is strongest where the water is deep, close to the undercut bank; it is slower where the water is shallow, and sediments settle out to form a bar, known as a *point bar,* against the slipoff bank; a

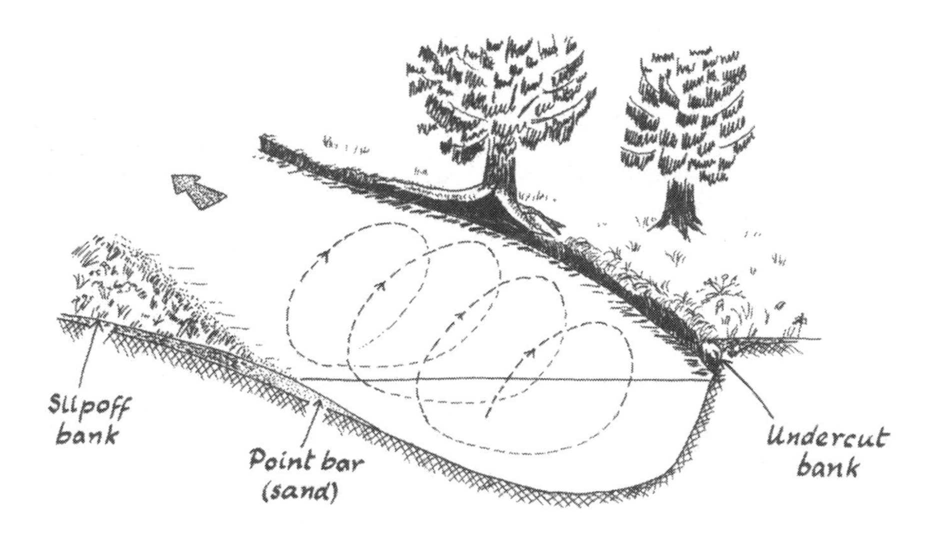

FIGURE 6.8. The flow (away from the viewer) of a river turning to the left as it rounds a meander bend. The dashed line with arrow heads shows the helical current; it descends as it approaches the undercut bank and ascends again, just above the sloping riverbed, as it approaches the slipoff bank. The current's horizontal direction is slightly to the right of the downstream direction near the surface, and slightly to the left near the bed. Compare figure 5.7.

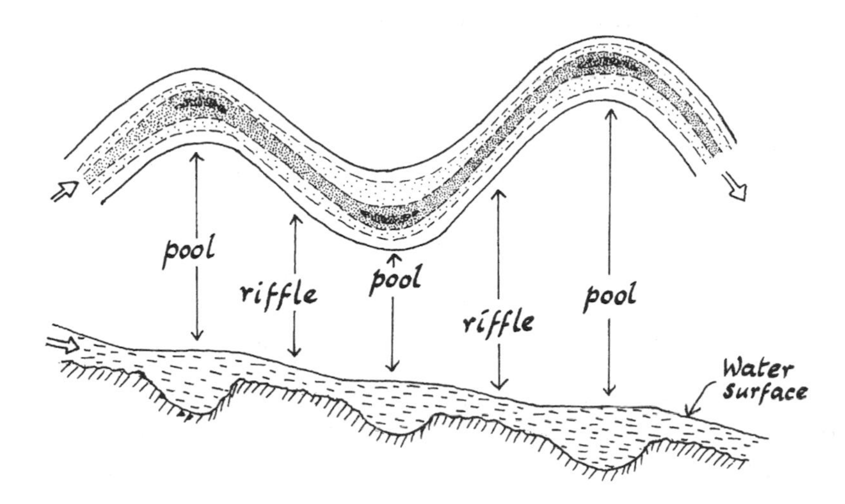

FIGURE 6.9. Map of a short section of a meandering river, and a longitudinal section along the thalweg (the deepest part of the channel). The map shows underwater contours, with deeper water shown by heavier stippling. Note that the deepest water (the pools) is near the outside of each curve. The section also shows how the slope of the water surface varies; it is almost level over the pools, and slopes more steeply over the shallows (the vertical scale is exaggerated for clarity).

point bar sometimes forms a long, narrow point (hence the name) of sand and gravel at the tip of a meander lobe.

Now consider the direction of the current as it rounds the bend of the meander. At the surface, the water strikes the undercut bank obliquely and is deflected, as we noted above. The deflection is not all in the horizontal plane, however; the current is directed downward as well, and flows back across the channel just above the bed. In short, the current follows a helical route, as figure 6.8 shows; compare the figure with figure 5.7 showing the paired helical currents that sometimes form in a straight stretch of river.

Maps, and cross-sections at right angles to the channel are still not enough to give a complete picture of the three-dimensional shape of a meandering river. To appreciate the shape fully, we must also look at a longitudinal section of the river, or its profile, as in figure 6.9.

As the figure shows, the channel is deeper at the bends than in the straight sections between bends. Pools and shallows alternate. Moreover, as the underwater contours on the map show, the depth varies across the channel as well as along it, with the deepest parts, the pools, being close to the outer side of the curves. As a result the course of the deepest "thread" of flowing water, a course known as the *thalweg,* is more zigzag than the center line of the channel.

The longitudinal section in the lower part of the figure is a section along the thalweg. The section shows the alternating pools and shallows, and how the water surface slopes (the vertical scale is greatly exaggerated to make the slope apparent). The slope is slightly steeper in the *reaches* (the straight segments of the river between bends) than in the pools.[21]

Thus there is more to a meandering river than a succession of regular curves making a bold pattern across a level landscape. Corresponding changes in the water's depth amount to "vertical meanders." The ups and downs of a river's bed are minuscule in scale compared with its curves in the horizontal plane, and are usually impossible to see through the muddy water of alluvial rivers in the lowlands. But inconspicuous or not, they are an integral part of a river's three-dimensional structure. Vertical meanders are most easily seen, and best appreciated, in clear streams in hilly country, streams that can be easily waded and examined close up. The alternation between deep, dark pools and bright, rippling shallows is one of the most attractive characteristics of such a stream, which is known as a pool-and-riffle stream. *Riffle* is a technical term in hydrology: it means a stretch of river where the bed consists of coarse sediments, and the water is shallow and fast-flowing; breaking wavelets often turn the surface into "white water."

Meandering rivers become less common, and pool-and-riffle streams more common, as you travel from alluvial lowlands up to higher ground. Inland of the alluvial lowlands, stream channels are in many places forced to follow routes controlled by the underlying bedrock. Streams having both meanders and a pool-and-riffle pattern, as in figure 6.9, are found in upland valleys that happen to have a wide, flat floor of alluvium, but elsewhere you are more likely to come across a pool-and-riffle stream whose bends are determined by the geometry of the bedrock rather than being true meanders. Pool-and-riffle sequences can develop where meanders cannot; in other words, vertical meanders can form where conditions prevent the appearance of horizontal meanders.

The pool-and-riffle pattern resembles an ordinary meander pattern in being constructed by the action of the stream itself. Visualize a stream in spate, fast-flowing, sediment-laden, and scouring its bed. The scouring will certainly be uneven and will produce extra deep hollows—in other words, pools—at soft spots. In doing this, the stream loses some of its energy and some of its load-carrying capacity; thereupon it begins to deposit some if its sediment, creating shallow, usually gravel-bottomed riffles downstream of each pool. As it dumps the heavy part of its load, the stream's energy is restored, and with it its power to scour out the next pool. And so the process continues. The principle of least action no doubt accounts for the details. Pool-and-riffle sequences are constructed at times of high flow, and remain unchanged when the flow slows down.

Vertical meanders cannot develop as regular a pattern as ordinary, horizontal meanders because their spacing is controlled to a large degree by the shape of the bedrock. In spite of this constraint they are not altogether irregular; the distance from one pool to the next is nearly always between three times and nine times the channel width,[22] with an average close to one-half that of the wavelength of horizontal meanders. This is what we should expect if the mechanics of meander formation are the same in the vertical as in the horizontal: as figure 6.9 shows, when the patterns are geometrically perfect, there are two pools in every S of the horizontal meanders.

Pools and riffles tend to be locked in place by the bedrock and cannot migrate downstream in the way meanders are apt to do. However, though riffles as a whole don't shift, the pebbles of which they are formed can migrate individually; when a stream is flowing fast, it often picks up pebbles from one riffle and deposits them on other riffles downstream.[23]

Pool-and-riffle streams are notable for the great variety of ecological habitats they provide for aquatic plants and animals.[24] Anglers are well aware of this. Deep pools are the preferred habitat of large fish. The pools need not be part of a pool-and-riffle sequence: pools in backwaters or behind log dams are equally good. The highest-quality pools, those with deep, slow-flowing water and overhanging banks, contain the greatest total mass of living fish because they are home to the "big ones." Small fish, especially young salmon and trout, are better off in lower-quality pools where they are at less risk of being eaten by the large fish.

Young fish also prefer riffles because that is where their small-sized food—aquatic insects and other invertebrates—is in greatest supply. Gravel-bottomed riffles are the chief spawning sites of salmon and trout. The eggs and, subsequently, the young fish can hide in the interstices among the stones, out of reach of predators; at the same time, good spawning gravels must be free of fine sand and silt, which can smother and suffocate baby fish. The swift current of riffles also brings in a constant supply of well-oxygenated water and carries away wastes. At times of low water, riffles are apt to dry up (their gravel bottoms turn into exposed bars), and then the pools serve as refuges for all aquatic life capable of moving and unable to endure drying.

If we follow a pool-and-riffle stream downstream, we are likely to find that it "grows" into a larger river meandering across an alluvial floodplain before it reaches the sea. Now suppose we had followed it upstream, into rugged mountains. The stream is small near its source, and its channel narrow, so narrow that a few large boulders suffice to dam it; a single boulder may be adequate if it is big enough. In terrain like this, pool-and-riffle channels are replaced by *pool-and-step* channels (figure 6.10).[25] Steps form where gravel and sand have been washed away from beneath big boulders, leaving them unsupported; the boulders then roll over and block the channel. You will often find several boulders combining to form a natural dam so well "constructed" that it looks artificial: the boulders have rolled against each other and happened to fit together, with one of them functioning as a keystone locking them all into a rigid structure.

Pools alternate with the steps; sometimes pool-and-step channels are so regular they look like a flight of stairs. In extremely rugged country, stepped channels are carved from bedrock rather than loose boulders. The steeper the channel, the steeper the "staircase" and the faster the current. As the current

FIGURE 6.10. A pool-and-step stream.

speeds up, more and more of the water flows at supercritical velocity (section 5.8); when more than half the surface water is flowing supercritically, the stream ranks as a *cascade,*[26] a staircase of small waterfalls, each one dropping into a *plunge pool.*

The plunge pool at the bottom of a waterfall is hollowed out by *cavitation,* an especially powerful form of erosion. What happens is this. The water speeds up going over the falls, causing its internal pressure to drop and some of the air dissolved in it to escape as bubbles: the water is therefore full of bubbles when it first comes to rest. The bubbles (cavities) then collapse, releasing shock waves strong enough to crack the rocky walls of the pool, and the rock fragments are washed away.[27] Cavitation is the fastest form of erosion; it happens only below waterfalls, because only there is the flow fast enough to produce the necessary bubbles in proximity to solid rock. As a plunge pool grows bigger, it undermines the cliff over which the falls are falling; sections of the cliff break off, and the lip of the falls retreats. All cascades and waterfalls are affected by cavitation; it is causing the Horseshoe Falls at Niagara to recede upstream at a rate that averages more than a meter per year.

6.7 The Flow below Ground

Huge volumes of fresh water flow unseen, below ground level. I've already mentioned (section 3.1) the underground streams and rivers in karst land, where the bedrock consists of cavernous limestone, and will return to the topic below. The flowing water in limestone caves account for only part of the total underground flow, however. Plenty of water flows below the surface in most river valleys.

Anybody who has gazed at a river flowing over a bed of gravel or cobbles must have pondered whether the river could be said to have a bottom strictly speaking; water is obviously trickling through the myriad interstices among the stones, as well as flowing over them, and it seems reasonable to assume that the river water merges into the groundwater below without any definite boundary.

The water is indeed continuous all the way down, except in a desert river, or a river that has carved its channel through massive, crack-free bedrock. In the majority of rivers, the water visible in the channel is directly linked to the invisible groundwater. But conditions change markedly going from one to the other. The obvious differences are that the visible water—the river in the ordinary sense—is comparatively fast-flowing and well-lighted, whereas the groundwater under it moves sluggishly, if at all, in impenetrable darkness. The transition is not abrupt, however. It takes place in what is known as the river's *hyporheic zone,* an important part of most rivers from the ecological point of view,[28] but one whose very existence most observers are unaware of.

The width of a river's hyporheic zone varies enormously from one river to another.[29] Normally it is only a little wider, if at all, than the channel itself, but occasionally it extends as far as 2 or 3 kilometers beyond the banks on either side, so that the hidden, hyporheic "river" is far wider than the visible river.

The most likely cause for a wide hyporheic zone is that the river's constantly shifting meanders left a network of abandoned channels in which subsequent floods deposited coarse, porous sediments. Soil, and a cover of vegetation, developed on the deposits over the centuries, but the interstices in the porous layer have not become clogged; they still carry large volumes of hyporheic water. Although the water is concealed, its presence is given away by the pattern of the vegetation: trees and shrubs requiring plenty of water, such as poplars and alders, grow in narrow bands above the buried channels (figure 6.11).

The stretches of a river where the hyporheic zone is wide are usually quite short; they occur like beads on a string[30] along the length of the river's valley. Judging whether the underground water at any particular spot in the neighborhood of a river is continuous with the water in the river channel, in other words, whether it is the river's own hyporheic flow, is straightforward; it is

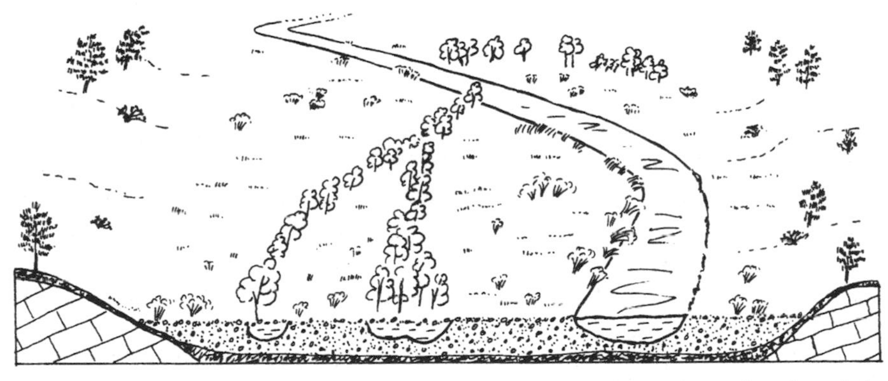

FIGURE 6.11. Bird's-eye view of a river floodplain. The bands of trees show where hyporheic channels flow below the surface of the floodplain.

done by repeatedly comparing water levels in wells drilled at various distances back from the banks with levels in the river itself; if the levels rise and fall simultaneously, it's safe to assume that the measurements come from a single body of water. Added evidence comes when the wells are found to contain stonefly larvae that can only have hatched from eggs laid in the river channel (more below on life in the hyporheic zone).

Hyporheic water differs in various ways from the river water above or beside it. It seldom flows at more than a few centimeters per hour; it contains less oxygen and more carbon dioxide than the water of the river proper; its temperature is less variable—cooler in summer and warmer in winter; and it contains more dissolved minerals. Hyporheic water differs, too, from the true groundwater below it, called *phreatic water;* this term is used when hyporheic water and phreatic water are not being lumped together (as they usually are) as "groundwater" in the broad sense. The change in physical characteristics may not be great, but there is often an abrupt change in the animals inhabiting the two kinds of water. Comparatively few species inhabit the phreatic zone, where the water is wholly unaffected by what happens up at the surface. Of the species that do, some live there permanently, in constant darkness, never straying into the light; they are mostly tiny, colorless, blind crustaceans, shaped like shrimps and pill bugs.[31] Others, for example, tiny water mites and seed shrimps, move up and down between the hyporheic and phreatic zones.[32]

Life in the hyporheic zone is far more varied than in the phreatic zone; figure 6.12 shows a few of the animals to be found there. Many live there for only part of their lives. Stoneflies, for instance, live there only as *naiads*—the immature forms of insects that spend their early lives in water; as adults they live in the air and lay their eggs in open river water. Other animals move up into the open channel and down into hyporheic water at will; the hyporheic zone provides a dry-season refuge for them if the channel dries up, or a wet-season refuge if the flow is damagingly strong when the river is in spate.

Because it is dark, the hyporheic zone is not an environment for green plants, or at any rate, not whole plants. The roots of many surface plants can easily reach down into hyporheic water, however, and when you find noticeably luxuriant vegetation on a floodplain, it is probably getting its nutrients from hyporheic water.

Hyporheic water could be said to form "streams" a short distance underground, but they are not independent streams; they are always in contact with surface waters directly above or beside them. True underground streams,

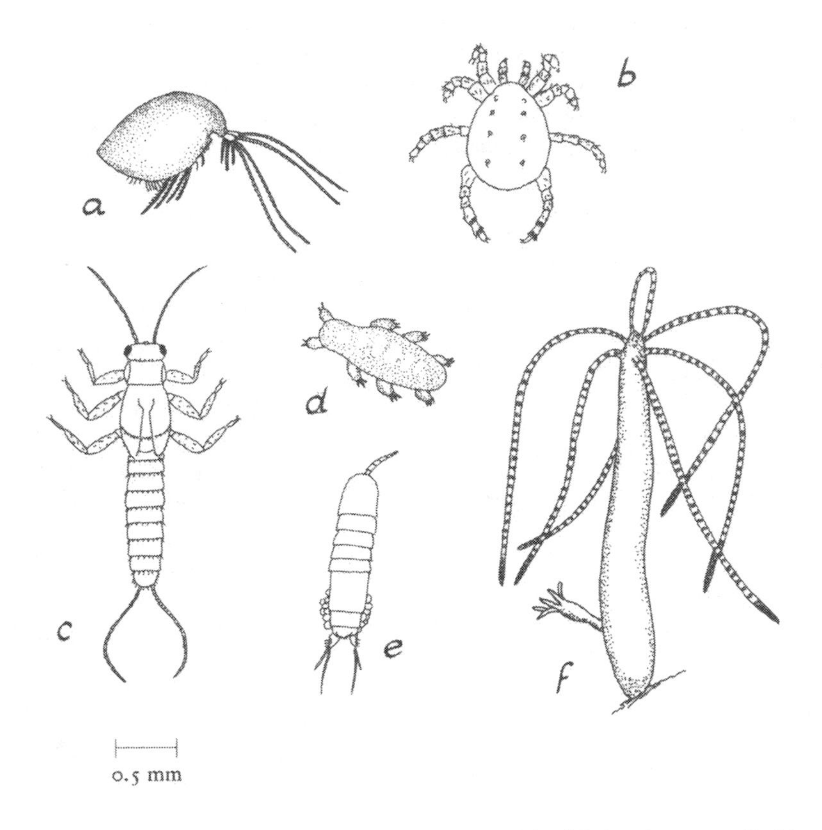

FIGURE 6.12. Some of the animals in the hyporheic zone. *(a)* Seed shrimp (or ostracod); *(b)* water mite; *(c)* stonefly naiad; *(d)* "bear animalcule" (or tardigrade); *(e)* copepod; *(f)* hydra.

flowing deep beneath the ground for great distances, exist too. They are found only in karst country (section 3.1), land where the bedrock consists of limestone pure enough and hard enough to be highly resistant to mechanical erosion. In karst the rock is chemically eroded instead; that is, it is gradually dissolved by mildly acidic rainwater trickling down through every available crack and fissure. Falling rain dissolves some of the carbon dioxide in the atmosphere, so that all rainwater is slightly acidic, even when it is free of the manmade atmospheric pollutants that produce "acid rain" in the pejorative sense.

Underground drainage systems in karst[33] are at least as variable as surface drainage systems and certainly more complicated, since they ramify through three dimensions; not only that, they are invisible, and difficult to map.

Some drainage systems in karst consist of intricate networks of small-bore *conduits* (the name for tunnels entirely filled by an underground stream); such a system is the underground equivalent of the upper reaches of a surface stream, where it is joined by numerous small tributaries; and some karst systems consist of large conduits with few tributaries, like the downstream reaches of a large river. The former kind of system, with a filigree of fine conduits, develops when the water dissolving the limestone is rainwater percolating directly down, through solution-enlarged holes and cracks, from the surface of the limestone bedrock. The latter kind of system, with a few large conduits, is usually the result of an already large stream reaching the karst region from somewhere else, after it has flowed a long way on the surface, over geologically different bedrock.

Whatever the subsurface pattern of the conduits, karst makes distinctive scenery (figure 6.13). Regardless of the climate, the land seems dry because incipient streams soon disappear underground via *swallow holes* (or *swallets*), vertical shafts penetrating deep into the rock; the entrance to a swallow hole may be at the bottom of a big, circular funnel, or it may be small and inconspicuous, concealed among grass and shrubs. Long, parallel-walled, vertical

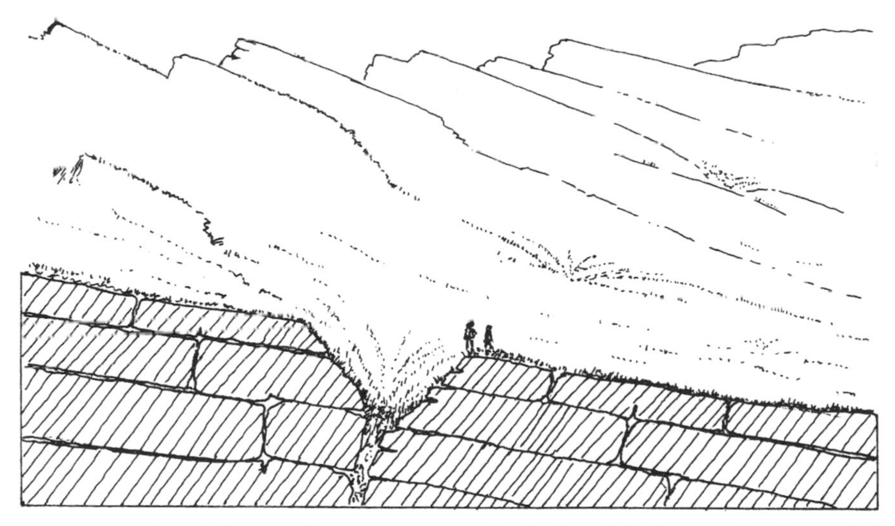

FIGURE 6.13. A typical scene in karst country. Note the dry funnels in the ground surface eroded by runoff into steep tunnels in the limestone.

clefts, known as *grikes,* provide another route to the underground world; the grikes are (fortunately) seldom wide enough to fall into, and their walls disappear into dark, invisible depths. Few ponds or lakes collect on the land's surface; in their place are dry hollows and dells where surface runoff has dissolved a funnel for itself down into the bedrock or where the surface has subsided because the ceiling of a cave has given way below.

To return to the underground streams and rivers themselves: they carry loads of sediment, which they pick up and deposit just as surface rivers do. They differ from surface rivers, however, in their behavior when in flood. A river confined to a conduit has no floodplain it can overflow onto horizontally. When a cave system floods, the overflow moves vertically, filling old, abandoned conduits at higher levels that may have been dry for many years. In some cave systems the water level can rise more than 100 meters in a few hours at times of extreme flooding.

6.8 How Plants and Animals Affect Streams

So far we have considered only the interactions between flowing water and the nonliving material—everything from clay to bedrock—that it touches. Flowing water molds this material, by eroding it, shifting it, undercutting it, or carrying and depositing it; at the same time the material molds the water by forcing it into meanders, or by accumulating in bars and other structures that deflect the water's flow.

In this section we consider the effects of the living world, specifically trees and beavers, on how streams flow and behave.

First, trees. In forested country there is more to the sediment load of a stream than clay, sand, and gravel; it obviously carries a load of organic sediment as well, chiefly of fallen leaves, twigs, and branches, some afloat and some waterlogged. Likewise with the obstructions to a stream's flow—logs are commoner than boulders. Pool-and-step channels are common in forest streams, but with fallen trees rather than interlocked boulders as the steps.

Pool-and-step streams look much the same regardless of whether the steps are made of logs or boulders; the two kinds of steps differ enormously, however, in their permanence. Whereas a boulder dam will last for centuries if its component rocks are firmly interlocked, a log dam, being organic, will decompose relatively quickly: as soon as it has been weakened and rotted by microbes, the wood becomes fodder for tiny aquatic beetles that chew it to pieces.[34]

Log dams are formed most often across a stream of medium size. In its upstream reaches, the same stream is likely to be in a valley so narrow that a fallen tree will make a bridge above it; downstream of the "log dam zone" the stream is so wide that fallen trees do not reach from bank to bank (figure 6.14). Where log dams are numerous, stream life is especially diverse, because of the contrasted habitats provided by the still pool behind each dam and the fast-flowing water below it.

Not all the dams in a woodland stream consist of solid logs; if the flow is

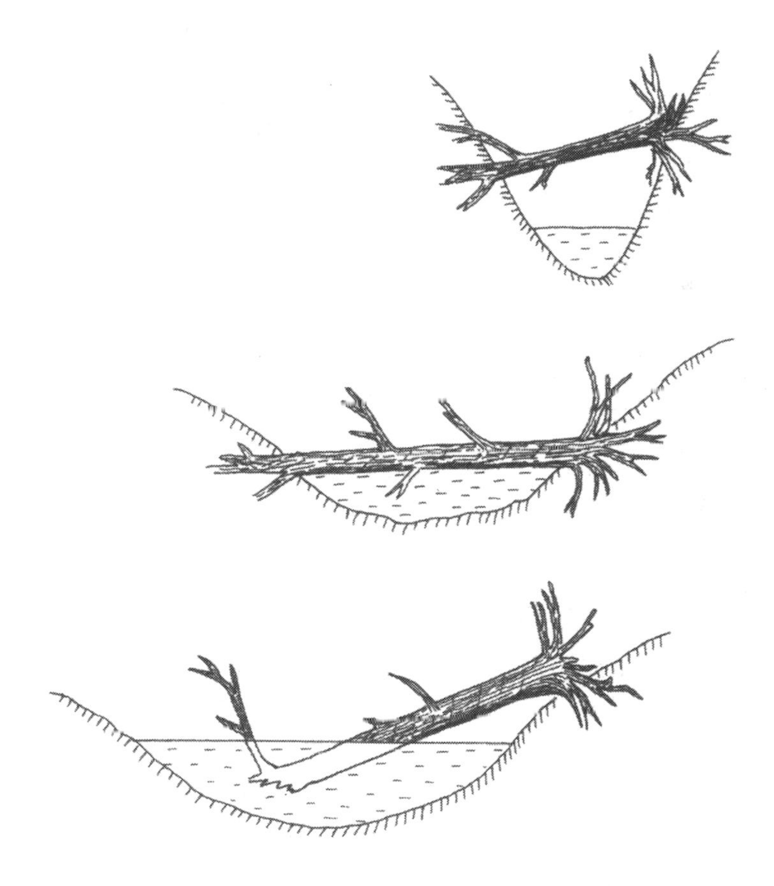

FIGURE 6.14. The resting position of a fallen tree across streams of different widths. The tree can make a bridge, a dam, or a partial obstruction.

not too fast, masses of debris and litter—branches, twigs, and leaves—often accumulate as dams. Such a dam acts as a strainer, trapping both woody debris and inorganic sediments (silt and sand) as they are brought down from upstream; the dam grows steadily bigger, until a chance flood washes it away. Debris dams are often spaced at fairly regular intervals, a few channel widths apart along the stream.[35]

A miniature working model of debris dams can nearly always be found if you explore forest trails after a rainstorm. A hard-packed, gently sloping trail among evergreen trees is the best place to look, where the ground is sprinkled with the needlelike leaves of such trees as pine, hemlock, or spruce. Rainwater dribbling down the trail gradually picks up needles until the load grows large enough to slow the water to a stop; the load is dumped as a mini-dam which strains out subsequent needles as they arrive, allowing the gently flowing streamlet (really only a trickle) to continue its journey unencumbered. It soon picks up a new load of needles, of course, and the whole process repeats. Low steps, consisting of bands of accumulated needles, cross the trail at intervals, which are often astonishingly regular, making a pattern easy to spot by anybody watching for it.

Next, consider beavers. Wherever they occur, their hydrological effects are profound. Obviously they create beaver ponds and cause the appearance of the different kinds of wetland that develop when the ponds are abandoned (for more on wetlands of all kinds, see chapter 10). It may be hard to believe, nowadays, that beavers' activities made much difference to the North American landscape as a whole, but that is because they are so much less numerous than they once were. Before Europeans invaded, the beaver population was probably between 60 and 400 million, spread over most of the continent; now it is probably less than 15 million, living in comparatively sparsely settled areas for the most part. The myriad forest openings created by abundant beavers have had a lasting effect; their influence persists for centuries.[36]

A tract of perfectly level land in a forest is nearly always the floodplain behind an old beaver dam. It is now an "island" of alluvium where a stream can flow unconstrained by bedrock and where it can carve small meanders exactly resembling (except for size) the sweeping meanders of a big river flowing across a wide coastal plain.

6.9 Deposition: Floodplains and Deltas

The sediment load of a river must eventually come to rest. As we saw earlier, the coarse fraction—gravel and coarse sand—which can be carried only by swiftly flowing water, drops out wherever the flow of the water is momentarily slowed, forming islands and bars in the upland stretches of the river. Some of the exceedingly fine fraction, consisting of clay particles so minute that they take months to settle in still water, is carried directly out to sea.

This leaves the middle-sized particles, mostly silt and fine sand, to consider; they are deposited when a river's flow slows down in the lowlands. The slowing down is caused by the gentler slope of the river's bed, the result of the deposition of sediment at times of low flow. It is a perfect example of positive feedback in action: the river slows, some of its sediment load is dumped on the riverbed, building up the bed so that it is more nearly horizontal, which slows the river still more, causing still finer sediment to be dumped . . . and so it goes on.

In the end the rising bed acts as a dam—in effect, the river builds its own dam—and next time the water level rises, it overtops the banks and floods the adjacent land. The floodwater flows at only a fraction of the speed of the river itself, and the sediment it carries is quickly deposited as a smooth layer of mud, in other words, as a bed of alluvium. Over the years, flood after flood deposits layer upon layer, and that is how floodplains are built.

Sometimes natural levees form: a *levee* is the raised rim of a floodplain, right next to the river's channel (figure 6.15), which builds up where a flood

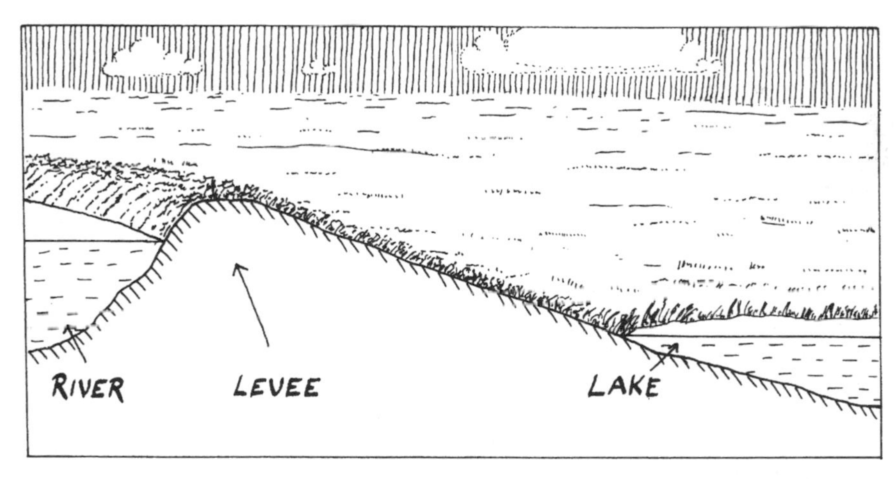

FIGURE 6.15. A natural levee. The vertical scale is exaggerated for clarity. The floodplain lake is below the level of the river surface. In populated regions natural levees are often artificially heightened, to protect the land behind them from repeated flooding.

ing river first overflows its banks; the sudden slowdown causes the water to drop the coarsest fraction of its sediment load immediately. As the residents of flood-prone regions well know, levees often build up on both banks to such a height that the level of the river between them is above the level of the floodplain on either side. When a high levee gives way, the resultant flood is especially catastrophic: water pours *down* from the river onto the land.

When it's not in flood, a floodplain is a passive scene. At low flow the river is confined by high banks made up of fine-grained, cohesive material. The river usually develops meanders, which, over the centuries, migrate endlessly, to left and right, and downstream. But however winding its course, an alluvial river usually carves for itself a channel with parallel banks and unvarying depth for a distance of many kilometers. If you pause to think about it, this fact is remarkable; it shows how disciplined flowing water can be when the land takes control.

At first thought, you might expect a river's floodplain to grow steadily higher as the years go by, but obviously it can't rise indefinitely. This raises the question, what becomes of the surplus sediment? The answer (provided the climatic and geological conditions remain unchanged) is that *all* alluvium is eventually washed out to sea. Meander loops are perpetually shifting this way and that across the floodplain; sooner or later every square meter will be undercut, so that any given chunk of alluvium will, for a while, become part of the sediment load again. It may be carried out to sea without further ado, or it may be redeposited and reeroded, perhaps many times over; in any case the sea is its ultimate destination.[37]

The site where a river deposits the last of its load—or almost the last, since a small fraction is carried far out to sea—is the point where it finally enters standing water and comes to a full stop. Even the fine clay, or most of it, settles out (unless the river flows into a freshwater lake), because the salts in sea water cause the clay particles to cohere in soft, loose masses that sink faster than separate particles. The deposited load is a *delta.* The majority of the world's great rivers have constructed deltas where they enter the sea. Ten familiar examples, listed in order of the amount of their annual discharge,[38] are the Amazon, Orinoco, Ganges, Mississippi, Irrawaddy, Mackenzie, Zambezi, Danube, Fraser, and Nile.

A delta is really a continuation of a floodplain, where excessive quantities of sediment have made the channel so shallow that it can no longer contain the river. The resultant floodwater drains through a number of separate channels (*distributaries*) that usually spread out fanwise, forking repeatedly, to form numerous outlets for the river. The distributaries are often very unequal in size, with one or two carrying the bulk of the flow while the rest are comparatively minor. Whenever a flood overflows from the distributaries, it deposits more sediment, some of it between the distributaries and some at their mouths. As they are repeatedly augmented, these deposits build up the delta's surface and extend it seaward, altering the shape of the coastline.

Where waves and strong tidal currents wash sediments away quickly, a delta doesn't project beyond the coastline or makes, at most, only a slight bulge in it; but where the sea's action is weak, the distributaries' banks and narrow floodplains grow out to sea as a bird's-foot delta, of which the Mississippi Delta is a famous example (figure 6.16).

The pattern of a delta's distributaries form a network rather than a tree. The radiating channels often come close to each other, and sometimes a pair unites to form a single channel; they may separate again downstream or unite with a third channel—the possibilities are unlimited. If a delta were to have a tree pattern, a canoeist paddling downstream via the distributaries could be sure, at every fork, that both branches led seaward. But since a delta's distributaries form a network, the seaward-bound canoeist can never tell, on reaching a fork, whether one of the branches (and if so, which one) doubles

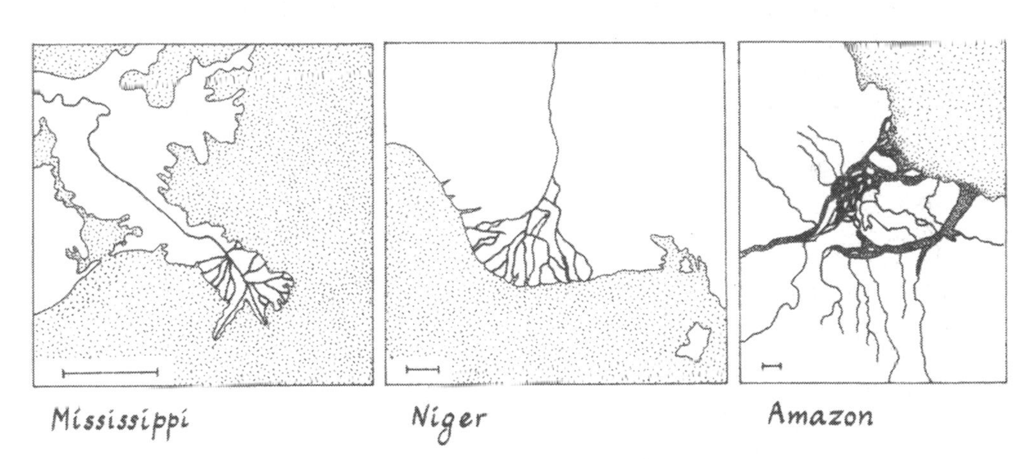

FIGURE 6.16. Three famous deltas, of notably different shapes. The Mississippi Delta is a bird's-foot delta. The Niger Delta is fan-shaped, because river-borne sediment has extended the coastline to seaward. The Amazon Delta consists of channels among the islands filling the estuary of the Amazon and its tributaries. Note the scale differences; the scale bar in each map represents 50 kilometers.

back toward land again. Only a very recent map would help, because distributaries never stop shifting, combining, recombining, and separating.

6.10 Floods: Good and Bad

Floods can be beneficial or disastrous. Much of the world's richest agricultural land consists of alluvial floodplain: without floods it would never have come into existence, and each successive flood, depositing a new layer of sediment, replenishes the soil's nutrient supply, renewing its fertility.

Disastrous floods are more newsworthy than good ones. Floods will continue to be disastrous as long as people choose to build their homes on floodplains. Sometimes the risk is worthwhile; it depends how often an area is likely to flood (more on flood frequencies in section 6.11).

Another form of "bad" flooding is that caused by dam building. One of the benefits an artificial dam provides is control—or, anyway, partial control—of downstream flooding (the other benefits are water storage and hydro power), but the water stored in the reservoir behind a dam is itself a flood, permanently submerging what was previously a tract of dry land and killing, or driving out, all its living occupants.

This is not the only damage done by controlling a river's flow. Many valley ecosystems *require* regular flooding; it is part of the natural timetable to which they have become adapted, and if the needed floods fail to materialize, profound ecological changes follow.

The effects of damming on the Peace-Athabasca Delta are an example.[39] They are, indeed, "a classic example of the kind of major ecological disturbance that can result when hydroelectric development is inadequately planned."[40]

The Peace-Athabasca Delta is in Canada's largest national park, Wood Buffalo National Park, established to protect several rare or endangered species, among them the whooping crane and the wood bison. It is an inland delta and the world's largest freshwater delta, with an area of nearly 4,000 square kilometers. Its shape is unusual for a delta because it is formed where four drainage systems converge: those of the Peace, Athabasca, and Birch Rivers and Lake Athabasca. The manner in which the delta floods every year is unusual, too. What happens is shown diagrammatically in figure 6.17.

The delta consists of all the wetlands and shallow lakes west of the western tip of Lake Athabasca, including Lake Claire; for most of the year the numerous channels crossing it drain northward, carrying water from the Athabasca

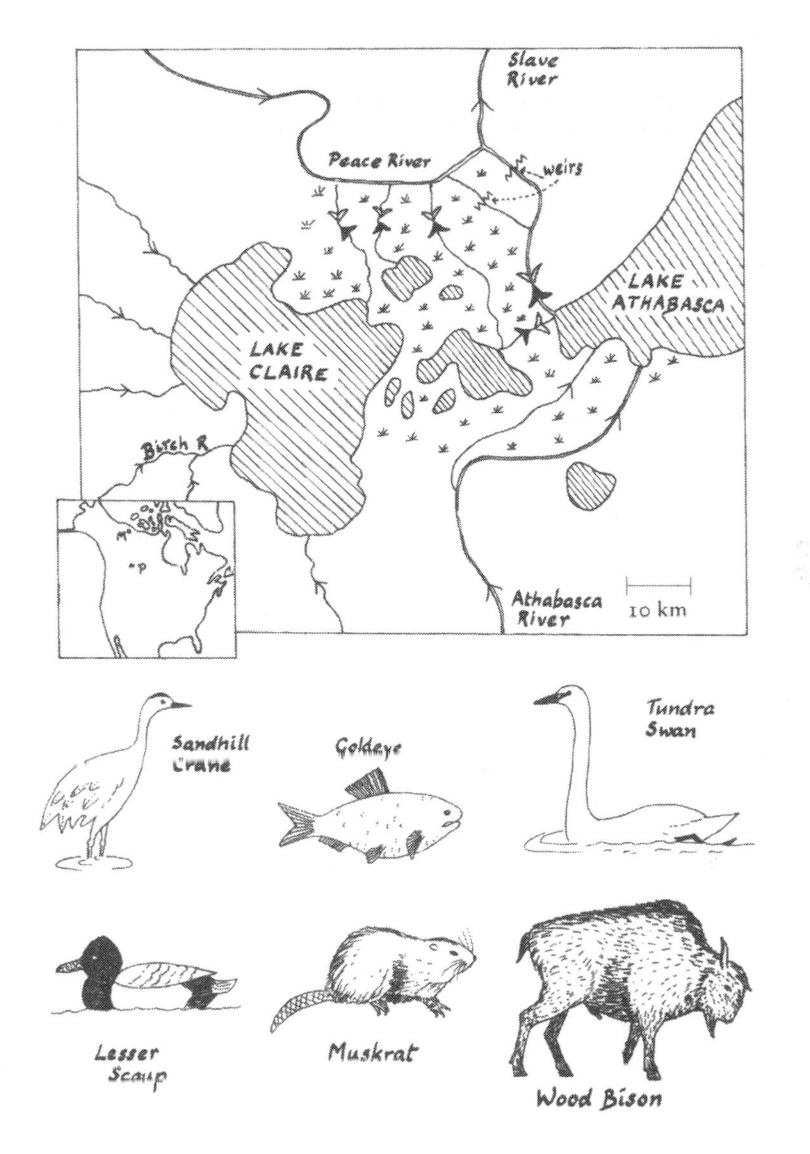

FIGURE 6.17. Map of the Peace-Athabasca Delta. Head-to-head arrows mark rivers whose flow reverses seasonally; black arrows show the normal direction, hollow arrows the direction when the Peace is in flood in spring. *Inset:* Map showing the locations of the Peace-Athabasca Delta *(P)* and the Mackenzie Delta *(M)*. The drawings (not to scale) show some of the animals whose habitat is threatened by the dam far upstream on the Peace.

River and Lake Athabasca north to the Peace River, which flows east into the Slave River. During this period the water level is higher in Lake Athabasca than it is in the Peace River. Conditions are different, however, in spring and early summer: then water levels rise everywhere because of spring runoff. The rise is particularly marked in the Peace River, which channels vast quantities of meltwater eastward from the distant Rocky Mountains, and for a period of 3 weeks or so the water level in the Peace is higher than in Lake Athabasca. The high water level in the Peace has been described as a "hydraulic dam," blocking the flow of the streams flowing north from the delta. They are not simply blocked, their flow is reversed: they flow south and flood the delta, submerging much of the land and refilling numerous ponds and lakes that would dry up were it not for this annual renewal.

This was the regular course of events until 1967, when a hydroelectric dam was built on the Peace, 1,100 kilometers upstream of its junction with the Slave. Once the Peace was "regulated," its annual peak levels were permanently lowered. Pronounced ecological effects gradually began to manifest themselves. The delta's extensive sedge and grass meadows, grazing lands for the world's largest unfenced bison herd, developed into willow scrub as the soil became drier. The delta ponds dried up, reducing the amount of shoreline available for waterfowl in an area that is one of North America's most important nesting and staging areas for waterfowl, a stopover for migrants traveling between the Pacific, Atlantic, and Gulf coasts and their northern breeding grounds. The big muskrat population was deprived of its freshwater habitat. Fish were affected, especially goldeye, which overwinter in the Peace and Slave Rivers and migrate into the delta to spawn.

The whole freshwater ecosystem is at risk from changes to water chemistry: the rivers flowing into Lake Claire from the west, and thence into the numerous small lakes linked to it, contain quantities of dissolved salts, which don't (normally) accumulate; instead, they are flushed out when the southward flow from the Peace floods the delta. If they are not flushed out, the ponds and lakes of the delta will eventually become saline, and encrustations of salt will cover the mudflats where shorebirds now feed.

Attempts have been made to mitigate these dire consequences of unwise dam building. Two weirs have been built (see figure 6.17) that have restored summer water levels to their natural height. But they will not restore the magnitude of the difference between high summer levels and low winter levels; and, as an evaluation has shown, "the weirs will not restore the biological communities to natural conditions."[41] In short, all is by no means well, in spite of attempted remedies.

Ecosystems like that of the Peace-Athabasca Delta are known as *pulsed ecosystems*. Their species and functioning require the pulsations of regular floods and cannot persist without them. Pulsed ecosystems are not limited to deltas; another typical site for them is the river valleys of the western prairies. Cottonwood (poplar) forests flourish, or used to flourish, on the valley floodplains; they were limited to the valleys because the high, grassy plains are too dry for trees.

Cottonwoods have very particular habitat requirements,[42] as we saw in section 4.5. Their seeds remain alive for less than a month after they are shed, and they cannot withstand competition from other plants; this means that the seeds must land on a bare, moist seedbed as soon as possible after they are shed; unless they do, they have no chance of germinating and becoming established. Prairie valleys provide ideal conditions, so long as they are not dammed. The rivers are very active at the time of the spring floods: their channels shift often, with the water uncovering previously submerged ground in some places and depositing fresh layers of sediment in others. The result of these "actions," performed by flowing water, is to expose tracts of wet sand and silt, devoid of competing plants, precisely when the newly shed cottonwood seeds need them. In a word, spring floods provide the regular "pulse" that keeps the cottonwood forests going. Without the floods new trees will not replace those that die of old age.

Damming a river stops the pulse and kills the downstream forest. The sediment required for seedbeds, as well as the water needed by seedlings, is held back by a dam. Instead of being carried to where it is needed, the river's sediment load is deposited in the still water behind the dam, creating a *silt shadow* downstream of the dam.[43] A silt shadow can be as much as 300 kilometers long in a slow, meandering prairie river.

The damage dams do to cottonwood forests have been observed in numerous places: in the Missouri valley in Montana and North Dakota (below the Garrison Dam), in the South Platte valley in Colorado, in the valleys of the Gila and Salt Rivers in Arizona, in the Bighorn River valley in Wyoming, in the valleys of the Waterton and St. Mary Rivers in Alberta, and in the Milk River valley in Montana (figure 6.18).[44]

6.11 Floods at Long Intervals: The Lottery

Up to this point, we have treated all floods as equal. In fact, of course, they vary enormously. Small floods (those in which the water only just tops a river's banks) are much commoner than big floods (those in which huge

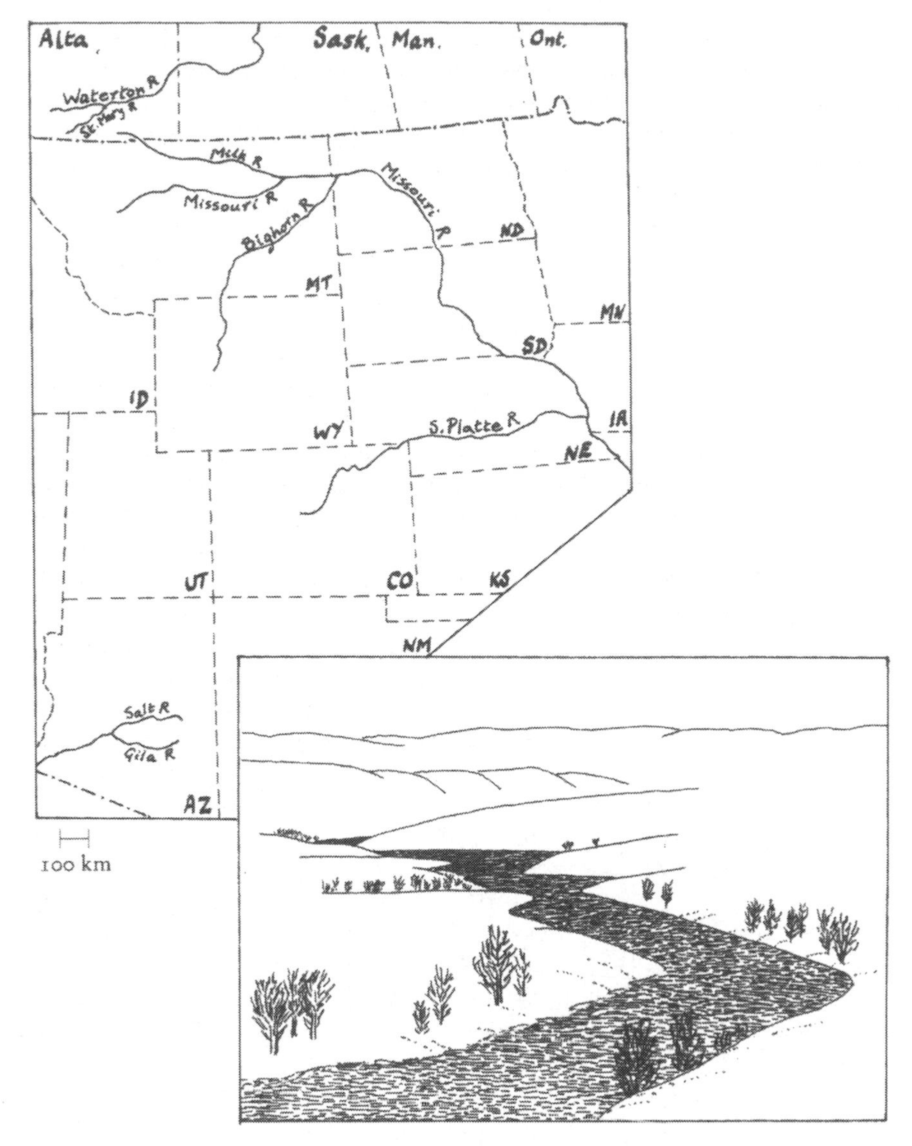

FIGURE 6.18. The map shows eight rivers where damming has injured the valley cottonwood forests downstream. *Inset:* Winter view of a prairie valley with patches of cottonwood forest.

volumes of water sweep over a vast tract of land). In other words, the chance of a big flood is always less than the chance of a small one, assuming predictions are being made months in advance, long before a river begins to rise.

We can be more precise than this: for any big river whose floods "matter" (meaning that they cause immediate economic loss), hydrologists have studied the river's natural flooding behavior; at an array of recording stations, records will have been kept for many years of the river's highest level, at that station, during the year. The peak flow—the moment when the level is highest—usually comes on about the same date at any given station; the date depends on the climate upstream of the station, on which months are the rainiest, and on when the runoff from melting snows reach peak values.

Given enough data, it becomes possible to describe the flood behavior at a station in terms of the *average return period* for floods of different heights.[45] For a given flood height this is the number of years, on average, between two successive floods equalling or exceeding that height; for example, one might find that floods of 3 meters or more come at intervals of 10 years, on average. Note the words *on average:* the average return period, whatever it is, tells nothing about actual return periods, which may have varied widely. Even the average is not an exact figure, but simply an estimate, based on records that may not go very far back into the past, and based on the dubious assumption that no long-term trends are affecting the climate or the river

All this imprecision means that the sizes of floods are impossible to predict until they have begun. If you are told that the average return period of, say, a flood of 3 meters or more at a location that interests you is 10 years, you can translate this to mean that in the light of experience, which may not be long, the chance that there will be a flood of that height (or higher) in the coming year is 1 in 10; if the average return period were 20 years, the chance would be 1 in 20; and so on. So whether you prepare for a big flood or not depends on your gambling proclivities.

The area submerged by a flood obviously depends on the flood's height: a big flood will spread more widely than a small one. On a floodplain where events take their natural course without human interference, there will be frequently flooded zones, occasionally flooded zones, and zones that are flooded only at still longer intervals. These zones are likely to have noticeably different vegetation; some ecosystems require infrequent floods—not too often and not too seldom—for their survival.

The white spruce woodlands just south of the arctic treeline on the Mackenzie Delta are an example[46] (the inset map in figure 6.17 shows the location). The woodlands grow on land where the average interval between floods is longer than 5 years but shorter than a century. Floods, when they come, submerge all the youngest spruce seedlings, coating their leaves with mud and preventing them from carrying on photosynthesis; therefore, in parts of the delta where floods come more often than about once in 5 years, the seedlings never have a chance to grow tall enough to overtop the floodwaters and survive (floods don't drown the seedlings, because they last for only a few days: it is the mud left on the leaves after the floods have subsided that does the damage).

The woodlands would not persist, however, were it not for occasional floods; spruce seeds require a seedbed of mineral soil, which floods provide by depositing new layers of sediment; the interval between floods must not be too long, however, or feather mosses growing on the woodland floor create a mat that covers the needed mineral seedbed. In a nutshell, the woodlands have a limited window of opportunity: excessively frequent floods kill new trees, and excessively infrequent floods prevent new trees from ever getting started; and without new trees a woodland will inevitably die out.

The frequency of floods in different zones of the Mackenzie Delta also controls the ecology of the numerous delta lakes that are periodically flood-filled: with no floods they would eventually dry out. The animals that depend on lakes need floods that are neither too frequent nor too rare. Annual flooding is too frequent; one sediment layer on top of another at annual intervals gives pondweeds and other water plants no chance to grow, and deprives muskrats and waterfowl of their usual food supply.[47] On the other hand, flooding must not be too infrequent; an arctic lake needs to be topped up by a flood every so often or its water level will fall to the point where the whole lake freezes to the bottom every winter,[48] with fatal consequences for many of its living inhabitants.

Humans are not the only organisms whose lives are affected by floods.

7

Lakes

7.1 Lakes: Where and Why?

In any memorable landscape a lake is often the principal item in the scene, the point to which the viewer's eyes are repeatedly drawn. After admiring the lake, however, an inquiring naturalist has two questions to consider and, if possible, answer.

The questions are inspired by two obvious truths: first, since a lake is a body of water that doesn't flow away, it must occupy a hollow surrounded on all sides by higher ground. The question is, how was the hollow, known as the *lake basin,* formed?

Second, a lake that doesn't dry up must be supplied with water from elsewhere or it would eventually evaporate, and it must have an outlet for surplus water, unless surpluses are always small enough to be disposed of by evaporation. The amount of water in the lake need not remain the same constantly. There may be a high-water, wet-season level, and a low-water, dry-season level; even so, the inputs and outputs obviously match each other well enough to keep the water level between these limits. The question is, where do the inputs enter, and where do the outputs leave?

In this section we consider how lake basins originate. A lake can't make its own basin, in the way a stream makes its own valley: the order has to be basin first, lake second. In broad terms lake basins are formed in three ways:

by movements of the earth's crust, by events that happened in the last ice age, and by more recent changes in the land surface that have happened since the end of the last ice age.

The oldest lake basins were formed by movements of the earth's crust. World-famous examples are the great rift valleys of East Africa, which contain numerous long, narrow lakes such as Lakes Tanganyika, Malawi, Turkana, Rukwa, and many others; the valleys—their lake basins—were formed around 10 or 12 million years ago. A rift valley is the trough that forms when the ground sinks between a pair of long, parallel faults in the earth's crust. Not all crustal ruptures and deformations are so large. Whenever the ground surface subsides over a fracture zone in bedrock (see figure 3.5), a long, narrow depression appears that resembles a miniature rift valley, and it is likely to hold a lake that resembles a rift valley lake except that it is a thousand times smaller.

The craters of inactive volcanoes often serve as lake basins. A well-known example is Oregon's Crater Lake (figure 7.1), whose basin is the huge crater (a *caldera*) left when a big volcano (in this case, Mount Mazama) blew up and then collapsed, 6,700 years ago; this isn't ancient in the geological sense; it is more recent than the last ice age, when a vast number of lake basins were formed.

Next consider lakes that owe their existence to the ice age. It has been rightly said that "most of the world's lakes are of glacial origin and occupy higher latitudes."[1] To see the evidence, look at a large-scale map of northeastern North America: it shows all the land north of the Great Lakes from the Atlantic coast to the western plains peppered with lakes of all shapes and sizes; their number seems uncountable. Nearly all are glacial lakes, the majority of them *ice-scour lakes*. They fill hollows that were ground out of the hard, ancient rock of the Canadian Shield[2] by the biggest ice sheet of the last ice age as it moved across the continent.[3] The ice sheet grew to be kilometers thick and had quantities of broken rock frozen into its lower surface; as it crept inexorably across the land, it functioned as a giant sheet of rough sandpaper scouring the rock below with unimaginable force. This is the way most of the innumerable lakes of "shield country" were formed.[4] They consist of bodies of clear water in basins of hard, clean rock, and a notable characteristic of a swarm of such lakes is that each one is at a slightly different elevation than all the others; this is difficult to detect in forested country but obvious in treeless tundra (figure 7.1). The reason the lakes are at different levels is that the bottom of the ice sheet wasn't flat and, therefore, the lake basins were created at different levels, independently of one another; given enough

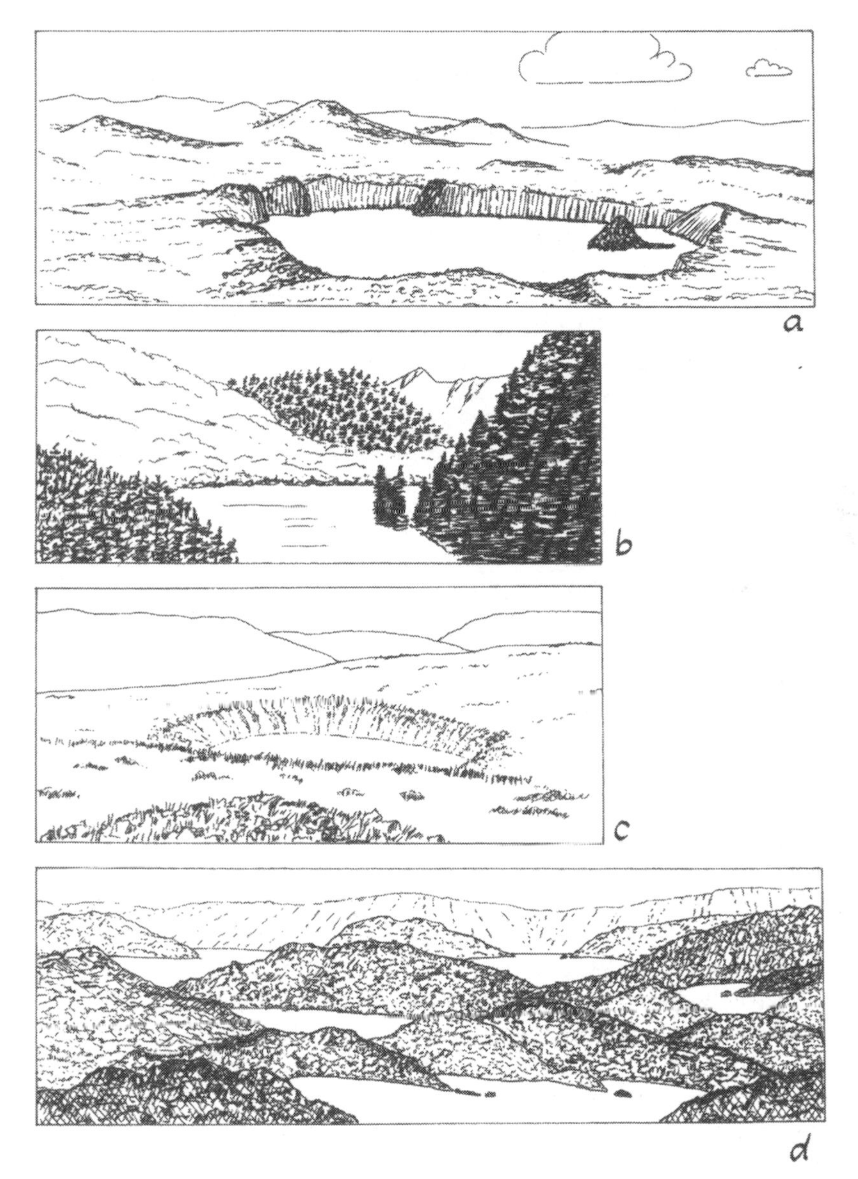

FIGURE 7.1. Lakes of different origin. *(a)* A crater lake; *(b)* a lake dammed by a landslide; *(c)* a kettlehole lake; *(d)* five ice-scour lakes, all at slightly different elevations.

time, stream erosion will mold the landscape and drain nearly all the small lakes into a few big ones, but because of the hardness of the shield rock, this will take many millions of years.[5]

Most of the many thousands of ice-scour lakes within the borders of the Canadian Shield are small, or medium-sized at most. Along the borders, however, is a group of five enormous ice-scour lakes—the Great Lakes of North America—constituting "the largest system of fresh, surface water on earth, containing roughly 18 percent of the world supply."[6] Other huge ice-scour lakes also lie on the border of the shield: the biggest are Great Bear Lake, Great Slave Lake, and Lake Winnipeg, respectively the sixth, eighth, and tenth biggest freshwater lakes in the world (figure 7.2).[7] Ice scour formed the basins of these lakes by deepening and enlarging preexisting river valleys. The lakes themselves began life as *ice-front lakes;* this happened when meltwater from the shrinking ice sheet became ponded between ice cliffs on one side and rising ground on the other.[8]

The ice age is responsible for another, quite different kind of lake. These are lakes whose basins are *kettleholes*. While an ice sheet was melting, it thawed most rapidly along its margin: thus the margin (not the ice itself) continually retreated as the ice sheet shrank. At the thawing "front," the decaying ice disintegrated into blocks, which became buried in sediments dropped by meltwater flowing from under the ice margin and swirling around them. Once buried, the blocks were insulated by the layers of sediment covering them, and their melting almost stopped; they were able to survive as underground ice blocks for thousands of years, shrinking at an imperceptibly slow rate. Melt they eventually did, however, leaving "sockets" or kettleholes, which became small, deep lake basins (figure 7.1). Kettlehole lakes and ice-scour lakes are easy to distinguish from each other: the former occupy hollows in sandy, muddy sediments containing loose pebbles and boulders; the latter are in hollows scoured in hard, solid rock.

The ice age was responsible for yet another form of lake basin. The melting ice sheet left extensive *moraines;* these are the heaps of rocky rubble, of boulders, pebbles, and clay, that the ice carried on its surface or frozen into its base, and that were left sitting there when the ice melted. This glacial junk has a notably bumpy surface, with numerous hollows ready to become lake basins. The 10,000 lakes for which Minnesota is famous occupy basins formed in this way.[9]

Lake basins continue to be formed by various agencies at the present day: landslides block river valleys (figure 7.1); hollows are left when sediments are deposited unevenly on floodplains; and shallow ponds form on the arctic

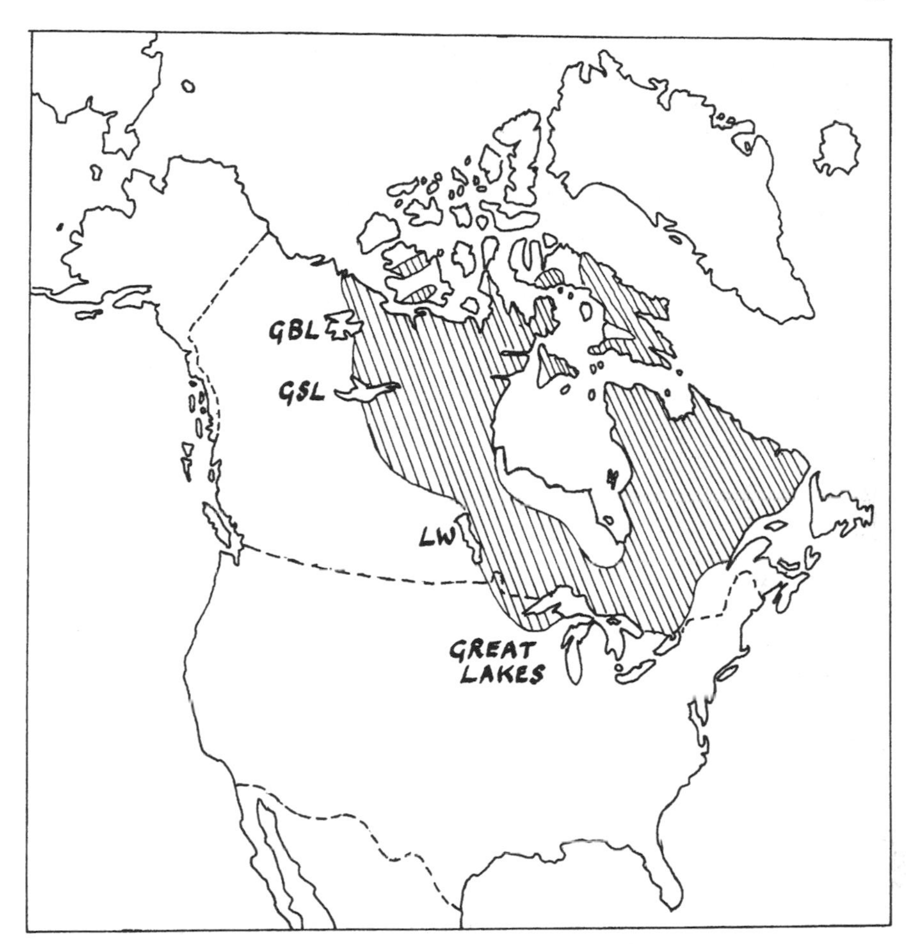

FIGURE 7.2. Map showing the biggest of the ice-scour lakes surrounding the Canadian Shield. The hatched area shows where the rocks of the shield are exposed at the surface (the shield as a whole is more extensive, but some of its periphery is buried under younger rocks). *GBL*, Great Bear Lake; *GSL*, Great Slave Lake; *LW*, Lake Winnipeg.

tundra[10] when the ground surface is deformed by repeated freezing and thawing. Any lake, wherever you find it, deserves to have its origin contemplated.

7.2 How Lake Water Comes and Goes

We come now to the second of the two questions mentioned above: where does the water in a lake come from, and where does surplus water leave it?

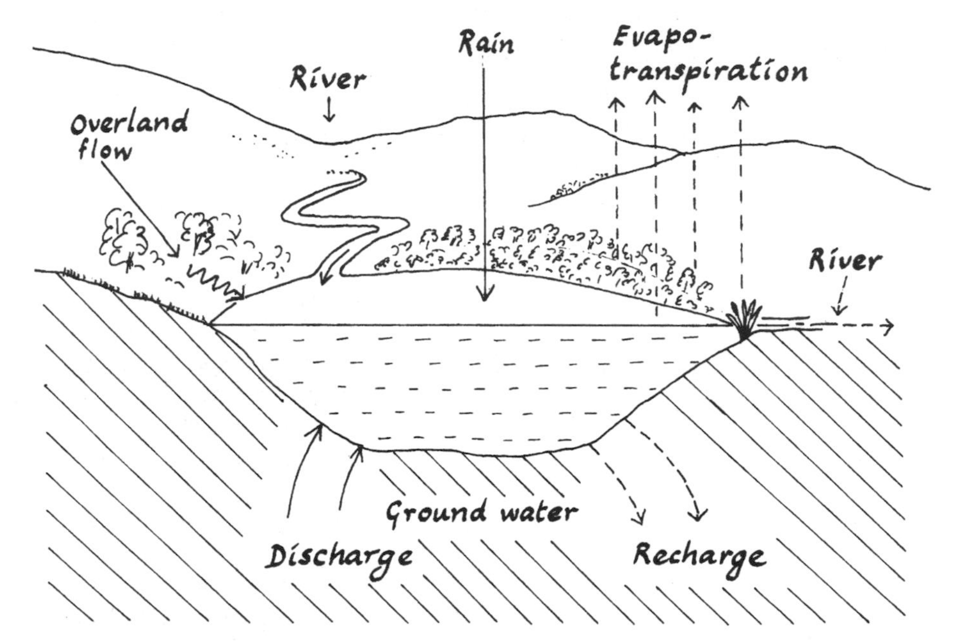

FIGURE 7.3. Diagram showing the inputs to (solid lines) and outputs from (dashed lines) a lake.

The answers, at least in simple form, can be discovered just by thinking about the questions. Water enters a lake from inflowing rivers, from underwater seeps and springs, from overland flow off the surrounding land, and from rain falling directly on the lake surface. Water leaves a lake via outflowing rivers, by soaking into the bed of the lake, and by evaporation. So much is obvious (figure 7.3).

The questions become more complicated when actual volumes of water are considered: how much water enters and leaves by each route? Discovering the inputs and outputs of rivers is a matter of measuring the discharges of every inflowing and outflowing stream and river, as described in section 5.4. Then exchanges with the atmosphere are calculated by finding the difference between the gains from rain, as measured (rather roughly) by rain gauges,[11] and the losses by evaporation, measured by methods described in chapter 12. For the majority of lakes, certainly those surrounded by forests, input from overland flow is too small to have a noticeable effect. Changes in lake level not explained by river flows plus exchanges with the atmosphere must be due to the net difference between what seeps in from the groundwater and what leaks into the groundwater. Note the word *net:* measuring the actual amounts of groundwater discharge (from the ground into the lake) and of groundwater recharge (from the lake into the ground) is a much more complicated matter than merely inferring their difference.

Once all this information has been gathered, it becomes possible to judge which are greater, the surface inputs and outputs or the underground inputs and outputs. If the former are greater, the lake is a *surface-water-dominated lake;* if the latter, it is a *seepage-dominated lake.* Occasionally, common sense tells you whether a lake or pond is surface-water-dominated or seepage-dominated. For example, a pond in hilly country that maintains a steady water level all through a dry summer in spite of having no streams flowing into it must obviously be seepage-dominated. Conversely, a pond with a stream flowing in one end and out the other, which dries up when the river dries up, is clearly surface-water-dominated.

By whatever means, a lake is constantly gaining water and losing water: its water doesn't just sit there, or, anyway, not for long. This raises the matter of a lake's *residence time,* otherwise called its *retention time* or *renewal time.* The residence time is the average length of time that any particular molecule of water remains in the lake, and is calculated by dividing the volume of water in the lake by the rate at which water leaves the lake. The residence time is an average; the time spent in the lake by a given molecule (if we could follow its fate) would depend on the route it took: it might flow through as part of the fastest, most direct current, or it might circle in a backwater for an indefinitely long time.

Residence times vary enormously. They range from a few days for small lakes up to several hundred years for large ones; Lake Tahoe, in California, has a residence time of 700 years.[12] The residence times for the Great Lakes, namely, Lakes Superior, Michigan, Huron, Erie, and Ontario, are, respectively, 190, 100, 22, 2.5, and 6 years.[13] Lake Erie's is the lowest: although its area is larger than Lake Ontario's, its volume is less than one-third as great because it is so shallow—less than 20 meters on average.

A given lake's residence time is by no means a fixed quantity. It depends on the rate at which water enters the lake, and that depends on the rainfall and the evaporation rate. Climatic change (the result of global warming?) is dramatically affecting the residence times of some lakes in northwestern

Ontario.[14] In the period 1970 to 1986, rainfall in the area decreased from 1,000 millimeters to 650 millimeters per annum, while above-average temperatures speeded up the evapotranspiration rate. The result has been that the residence time of one of the lakes increased from 5 to 18 years during the study period. The slowing down of water renewal leads to a chain of further consequences; it causes dissolved chemicals to become increasingly concentrated, and this, in turn, has a marked effect on all living things in the lake.

Changed residence times for lakes is one of the enormous web of consequences stemming from global warming.

7.3 Lake Structure: Layers upon Layers

Lake water is seldom at the same temperature from top to bottom. Most of the time, some of the water is warm and some cool; consequently, since the density of water depends on its temperature, the water arranges itself in layers: the least dense water forms the topmost layer, with successively denser layers at increasing depths.

Provided none of the water is colder than 4°C, this simply means that the warmest water is at the top, and the coolest at the bottom. This is because at temperatures above 4° the warmer the water the lower its density.[15] Below 4°, however, an extraordinary thing happens: the behavior of water changes radically—instead of becoming more dense as it cools, it becomes less dense. No other liquid behaves in this way.

The reason for the switch in behavior at 4° is this. In warmer water the molecules are in constant, random motion, like a dancing cloud of midges on a summer evening; the higher the temperature the faster the molecules move and the more space they collectively occupy; this is why the density of water becomes less as its temperature goes up. Conversely, the lower the temperature the more confined the space occupied by the moving molecules, but only until 4° is reached. At this stage incipient crystallization begins, and the molecules can no longer move freely: they are kept at "arm's length" from one another. This means that they occupy more space than they did when they were free to pack together and, consequently, the water becomes less dense. As the temperature continues to fall, molecular movement becomes more and more restricted, and the density less and less until, at 0°C, the water freezes solid.

Now consider the consequences of all this. Unless it is all of the same density, the water in a lake must become sorted into layers, with the least dense water at the top and the most dense at the bottom. The process is known as *stratification* (figure 7.4). If all the water in the lake is warmer than 4°, then the warmest water will be at the top and the coldest at the bottom. When all the water is colder than 4°, the coldest water is at the top (perhaps under a layer of ice) and the warmest at the bottom.

Next visualize what would happen if the temperature range were to bracket 4°; for example, suppose it were to range from 1° to 12°. Think about

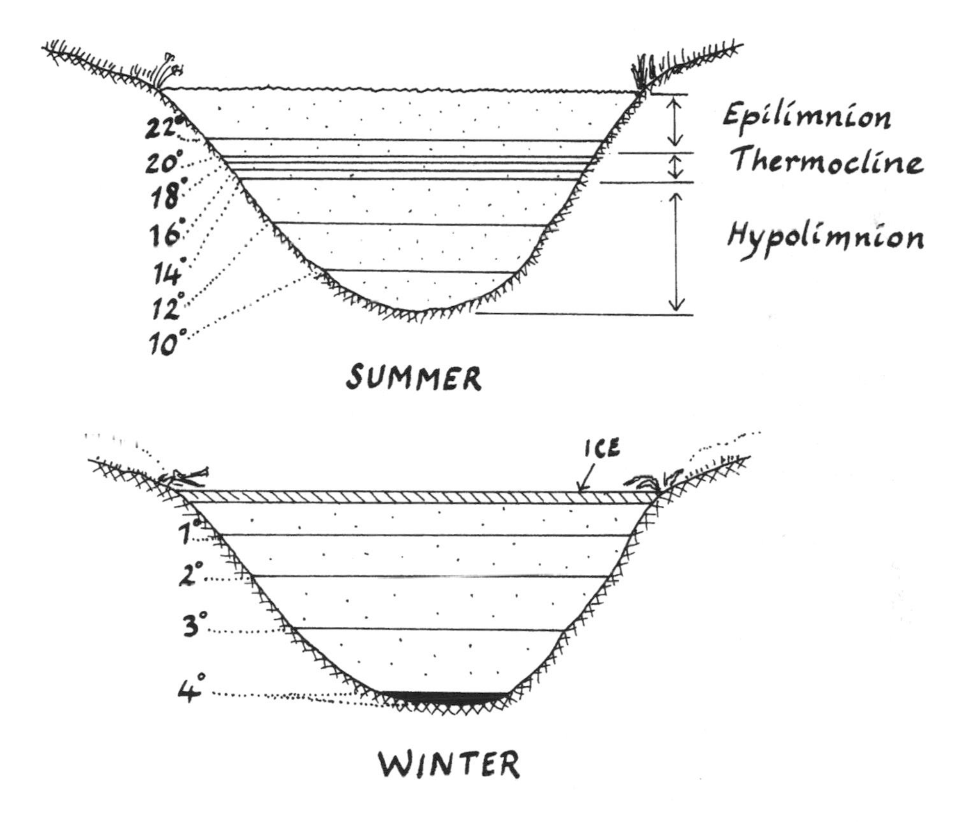

FIGURE 7.4. A lake in summer and in winter. The horizontal lines are isotherms, labeled with temperatures in degrees Celsius. Note that in winter the bottom layer is 4°C (shown black) throughout. In both seasons the lake is stratified, with density increasing from top to bottom, but the stratification in winter is called inverse stratification because the warmest water is at the bottom.

this, and you will see that it could never, in fact, happen: the densest layer, right at the bottom of the lake, would have a temperature of 4°. In each layer above, water at two different temperatures, one below 4° and one above 4°, would collect at the same level, and they would immediately mix. For example, water at 1° has the same density as water at 7° and they mix without stirring; the same goes for water at 2° and 6°; and so on. If the temperatures of all the mixed layers were above 4°, the lake would wind up as a "warm" lake as in figure 7.4a; this is how a lake is stratified in summer. But if the mixed layers were all cooler than 4° (as might happen if the starting temperatures ranged from 1° to only 5°, for instance), the lake would be a "cool" lake, as in figure 7.4b; this is the winter pattern.

Now look at the isotherms—the lines joining points of equal temperature—in the "warm" lake (figure 7.4a). Notice that they are not evenly spaced. On the contrary, they are closely bunched in a zone some distance below the surface, showing that the temperature drops rapidly as you go down through the zone. The zone is called the *thermocline* (or sometimes the *metalimnion*); its position varies greatly from lake to lake, but the top is usually somewhere between 5 and 25 meters below the surface. The warmer water above the thermocline is the *epilimnion* and the cooler water below it, the *hypolimnion.*

Within the epilimnion, water temperature is only slightly lower at the bottom than at the top—the isotherms are widely spaced (but even so the slight temperature drop below the surface can often felt by swimmers). Most of the time, this uppermost layer is well mixed. Currents and turbulence caused by the wind keep the water circulating, and it is this characteristic of the layer that defines the epilimnion: the bottom of the epilimnion is the level where mixing stops. Below this level the water is placid; marked stratification, with a pronounced temperature drop, becomes possible, producing the thermocline.

Below the thermocline is the hypolimnion, where the water is calm and cold; its temperature continues to drop, less rapidly than in the thermocline, right down to the bottom.

Conditions in the "cool" lake are entirely different. The temperature range is slight—the water cannot be colder than freezing at the top, nor warmer than 4° at the bottom; and there is no thermocline.

"Warm" and "cool" lakes don't stay that way permanently, of course. In temperate latitudes most lakes are warm in summer and cool in winter. As a lake changes from warm to cool in the fall, and back from cool to warm in the spring, a stage is reached at which the temperature is almost the same throughout; stratification disappears.

It now becomes much easier for the waters of the lake to circulate. It takes considerable energy to disrupt the stratification of a strongly layered lake, with pronounced density differences; but when the density differences disappear, as they do in spring and fall, nothing prevents free movement of the water up and down. Currents set in motion by the wind stir and mix the water; what follows is known as lake *overturn*. During overturn the whole body of lake water becomes thoroughly mixed; dissolved nutrients, dissolved oxygen, and floating microorganisms, which had all been stratified, are mixed too (see section 7.7).

Spring overturn and *fall overturn* are regular seasonal events in many lakes, but this doesn't mean that lakes are immune to overturn at other times, or that all lakes become stratified. Lakes shallower than 10 meters in windy sites are seldom stratified. In general, the deeper the lake, and the more sheltered it is from the wind, the more readily it stratifies and the longer the stratification lasts.

The amount of wind energy needed to cause overturn depends on the strength of the stratification—its resistance to overturn—and that depends on the magnitude of the density differences. The density differences in a warm lake are much greater than in a cool one,[16] and so a warm lake is less easily overturned than a cool one. This means that a comparatively gentle breeze can trigger spring overturn; fall overturn needs a stronger push.

Lakes in regions with mild winters may be unstratified through the winter; they become stratified in spring and remain stratified until the fall overturn which, for them, is the only overturn of the year.

Rarely, a lake that is both very warm at the top and very deep remains stratified for years on end. The hypolimnion of such a lake becomes stagnant and lifeless.

7.4 Movement in "Still" Waters: Waves, Ripples, and Seiches

We are used to thinking of the earth's fresh surface water as being either moving (streams and rivers) or still (ponds and lakes). "Still" water is seldom absolutely still, however. It moves in a variety of ways, one of which, lake overturn, we have already discussed. The movements are not part of the hydrological cycle; they take place within lakes, and some of them are tiny. But many have visible effects; they cause changes in a lake's appearance that draw the attention of lake watchers, and therefore they invite explanation.

First, consider waves: visualize the choppy surface of a windswept lake, with the waves scudding before the wind. Obviously the water cannot be shifting rapidly downwind as a whole; if it were, any floating object, from a log to a ship, would hurtle downwind at the same speed as the waves, like a surfer. In fact, if you watch a floating log, it seems merely to bob up and down as the waves pass beneath it.

The details of what happens are shown in figure 7.5; the log is seen end-on; the waves are advancing from left to right; the horizontal line in each

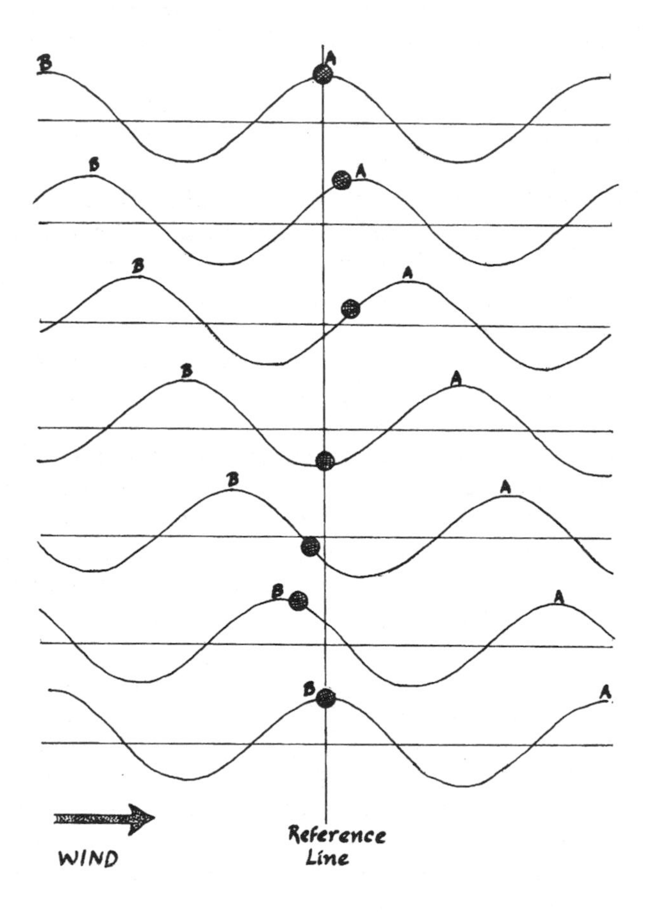

FIGURE 7.5. The movement of waves: each wavy line, from top to bottom, shows a slice of the water surface at seven successive times. The small cross-hatched circles show the end of a floating log. See text for details.

panel is the average water level. The successive panels from top to bottom of the figure show the scene at a succession of times. Suppose the log is at the crest of wave A when observation starts. As crest A moves to the right, the log slides down the back of it until it is in the trough between crests A and B. Then the advancing crest B buoys the log up again until, for a moment, it is at the crest of wave B, but as soon as that wave passes the log again slides down the back of the retreating wave, until the next wave comes along and buoys it up again; and so on, indefinitely. This explains the log's up-and-down motion. But notice that there is also a to-and-fro motion: the vertical reference line in the figure is an unmoving position marker, and the log moves first to the right of it and then to the left. Further examination of the figure reveals more: if you note the successive positions of the log relative to the intersection of the reference line and the average water level line in each panel, you will see the log's path is a clockwise circle. (If the waves moved from right to left, the path would be a counterclockwise circle.)

To see all this in real life, observe a wood chip floating on a lake whose surface is choppy with small, nonbreaking waves. Hold a pencil upright, at arm's length, between your eye and the chip, to represent the vertical line in figure 7.5, and take care to keep its position fixed; now concentrate on the chip and you will *see* that, rather than merely bobbing up and down, it moves in vertical circles, or, at any rate, in closed loops (because real waves are seldom as regular as those in the figure, the "circles" are seldom perfect). The chip is left behind by the passing waves and seems to be going nowhere. In actual fact the surface water and anything floating on it do drift downwind very slowly, because the wind always causes a slight current in addition to waves. The drift, known as *Stokes drift,* moves at about three one-thousandths of the wind speed.[17]

The motion of the floating chip must obviously match the motion of the particular little patch of water it is in contact with. It follows that the water itself is moving in vertical circles. It does so only close to the surface, however. The movement dies away quickly with increasing depth, so that at a depth of only one wavelength (the crest-to-crest distance) below the surface of choppy water, the rocking motion due to the waves is less than two one-thousandths of what it is at the surface.[18] Fish have no problem finding calm water, however big the waves.

"Ordinary" waves on a lake are very much a surface phenomenon (unlike internal waves, deep in the water, which are considered below). This raises the thorny question: given an absolutely calm surface, how do waves get started when the wind rises? It is obvious how rising winds enlarge the waves once they are already there, and obvious how a sudden downdraft slapping a small area of flat water on a generally calm day can create a patch of little wavelets (a *catspaw*), but it is not obvious how waves start when a horizontal wind blows across water that is absolutely flat, with nothing protruding for the wind to push against. The explanation is believed to be that minute fluctuations in air pressure dimple the water surface so that it never is perfectly flat: there always are imperceptible slopes that the wind can act upon, enabling it to *drag* the water.

When wind drag is gentle, it causes slow, shallow surface currents to flow, and tiny ripples to form. Drag always affects water in both these ways: a current flows, and waves or ripples corrugate the surface. The weak current produced by mild drag acting on smooth water is easily observable. To see it, look at a small, sunlit pond while a gentle breeze is blowing, and you will see that the particles—dust and pollen—suspended in the very topmost layer are drifting in a slight current, whereas those only a short distance below are stationary. Bright sun is necessary to make the submerged particles visible.

The smallest surface corrugations are *ripples,* which are not simply tiny waves: they differ from waves in kind as well as in degree. Consider the forces tending to flatten corrugated water: gravity pulls down the crests, and at the same time surface tension tends to make the surface area of the water as small as possible, in other words, to level it. Which of these two forces predominates depends on the wavelength of the corrugations. At a wavelength exceeding 2 centimeters, gravity is stronger than surface tension, whose effect on what one would think of as ordinary waves is negligibly small. But at wavelengths of less than 2 centimeters, surface tension becomes more important than gravity in restoring surface flatness; when this is the case the "waves" are technically ripples.

Strengthening winds cause currents to speed up and waves to grow larger. Storm rain boosts the effect of the wind appreciably.[19] Bulletlike raindrops, striking the waves' sloping sides almost horizontally, impart their energy to the waves and drive them forward. The raindrops also share some of their momentum with the air, so that the normal reduction in wind speed close to the surface is less than it would be if it were not raining.

Lake waves are never as big as ocean waves. The height to which waves grow depends on three things: the strength of the wind, the length of time it has been blowing, and the *fetch* (the distance the wind has come over open water). It is the last of these three—the fetch—that limits the height of lake waves. Here are a few numbers.[20] Imagine a gale-force wind (55 kilometers per hour) blowing across a deep lake for as long as it takes to raise a maximum "sea." If the fetch were 10 kilometers, the average wave height (crest to trough) would be only 66 centimeters, and the gale would have to blow for at least 1 hour for the waves to grow to this size. If the fetch were 50 kilometers, average wave height would reach 2.3 meters, provided the gale lasted 4 hours or more. With a fetch of 500 kilometers—attainable only along the length of Lake Superior—average wave height would be 2.7 meters, provided the gale continued for at least 25 hours. These wave heights are averages; some individual waves will be much higher; for the Lake Superior example (same wind speed and duration), about 10 percent of the waves would be higher than 5.5 meters; that is a tremendous height for a wave on a lake, to be expected only in the most exceptional circumstances.

The behavior of a big wave when it reaches the shore is familiar to everybody: the wave cannot retain its shape as it moves up the beach because the forward speed of the crest comes to exceed the speed in the trough ahead of it, which is slowed or reversed by the backwash of the preceding wave; so it rears up until the top topples forward, and finishes its shoreward journey as a foaming breaker. In technical terms, the water performs a hydraulic jump (see section 5.8).

Wind drag does more than build waves; it also drags water along bodily. After many hours of steady wind, the water level may be considerably higher at the downwind shore of a lake than it is upwind. The phenomenon is called *wind setup*. The surface of a big lake can also develop a slope because of strongly contrasted barometric pressures at opposite ends of the lake.

When the cause of the slope, whatever it is, disappears, levelness is restored—usually. It occasionally happens, however, that instead of simply returning to the horizontal, the water overshoots and piles up against the opposite shore; it then tries again to return to the horizontal, but overshoots again, and again and again. In a word, it sloshes from end to end (or side to side) exactly as the water in a wide shallow bowl does when you carry it. The physics is the same in both cases: sloshing sets in if the time taken for the water to flow from one end to the other and back in its container, whether bowl or lake basin, happens to coincide with—or resonate with—its natural period of free oscillation. In the case of a bowl of water, you can find the period of free oscillation by jolting the bowl and timing the resultant rocking motion of the water. If you then carry the bowl some distance, you can cause or inhibit sloshing by adjusting the speed of your steps so that the resultant jolting either reinforces the sloshing or damps it.

When a lake sloshes, the water movement is known as a *seiche*. The largest seiches happen in long, narrow lakes. They are imperceptible to the casual observer because the change in water level is usually slight. To recognize a seiche with certainty, instruments must be used, recording how the depth of the water changes as time passes. If the water level goes up and down at regular intervals, a seiche is in progress.

In fact, a seiche is an extremely long wave, one with a wavelength double the lake's length.[21] It behaves differently, however, from the waves we have considered hitherto, which move forward before the wind and are known as *progressive waves.* A seiche is a *stationary wave,* in which the water simply rises and falls in one place. In effect, a stationary wave forms when a progressive wave is reflected back on itself from one end of the lake to the other. The original wave and its reflection unite, becoming added to each other where both are up and canceling each other where one is up and the other down. Figure 7.6 shows what happens diagrammatically. The five panels show the "lay of the water" at a sequence of times, starting from the top. In each panel the thick line shows the position of the water surface; the vertical borders to left and right are the lakeshores, shown as vertical cliffs for simplicity. The dashed lines show the component waves—each a reflection of the other. When both act together (top and bottom panels), the water reaches its highest level at one end of the lake and falls to its lowest at the other. When the component waves oppose each other exactly (middle panel), the water is horizontal. At intermediate stages (second and fourth panels) the component waves combine and cancel to an intermediate degree. The result is that the greatest up and down movements happen at opposite shores, while the water level is motionless halfway between. Notice, too, the horizontal movements of the water, shown by the arrows in the figure. It travels back and forth at varying speeds, shown by the arrows' lengths. When the water level is at its highest at one end of the lake, the water is slack—motionless; then it begins to move, with increasing speed, toward the other end. The flow is most rapid when the surface is level, whereafter it decelerates as the crest builds up at the opposite end. The foregoing describes the physics of sloshing as it can easily be seen happening in a handheld bowl.

The period of a seiche—the time required for a complete cycle—depends on the shape of the lake basin, especially its depth. In general, the shallower the water, the longer the period; this is because a moving wave "feels" the bottom (is slowed by its resistance) if the depth is less than half the wavelength, which is always true for seiches because of their tremendously long wavelengths. Here are some representative periods:[22] in Lake Huron (length

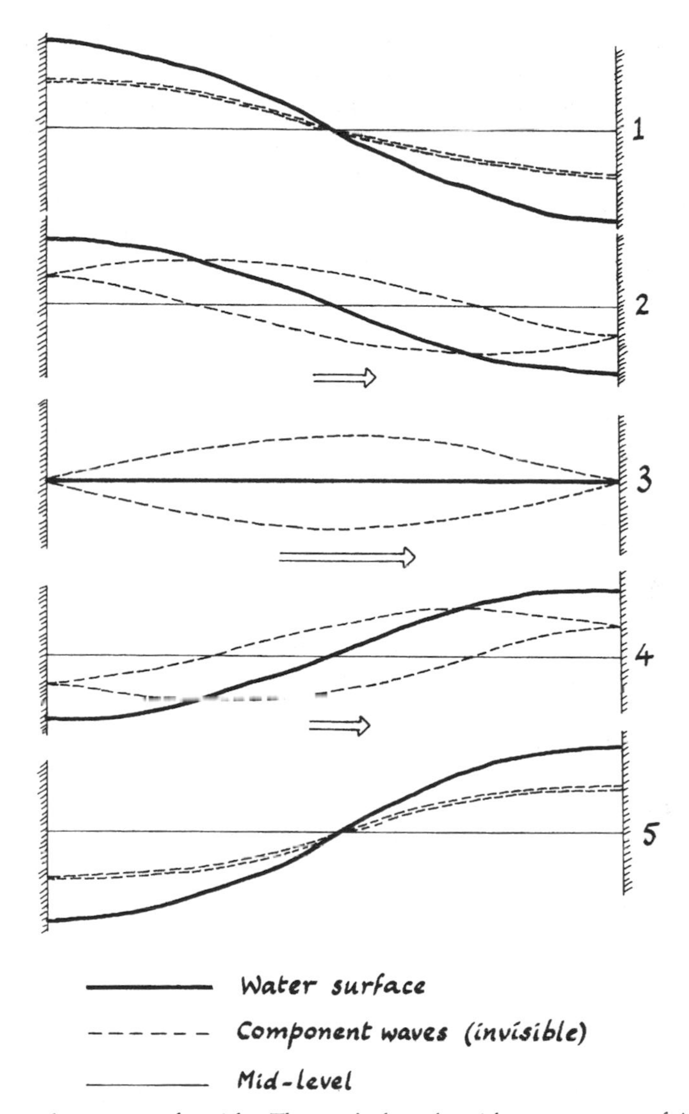

FIGURE 7.6. The progress of a seiche. The panels show the seiche at a sequence of times, numbered 1 to 5, from the beginning to halfway through one cycle; in the second half-cycle the water will sink on the right and rise again on the left. See text for details. Note that the vertical scale is about 10 million times the horizontal scale.

444 kilometers and average depth 76 meters) the period is a bit less than 7 hours; in Lake Erie, which is not much shorter than Huron (400 kilometers) but is much shallower (21 meters), the period is twice as long (13 hours). For comparison with these huge lakes, consider Lake Mendota in Wisconsin, a lake whose behavior has been closely studied; its length is 9 kilometers, its depth about 13 meters, and the period of its seiches only 26 minutes.

The height to which the water level rises in a seiche—its amplitude—depends on the strength of the forces that caused it. In a small lake like Lake Mendota, the amplitude of a typical seiche is only 1 or 2 millimeters. Some seiches in Lake Erie have amplitudes up to 2.4 meters.[23] Seiches are not spectacular enough to be distinguishable from water-level changes due to other causes, such as ordinary waves or wind setups that don't start seiches. Once the original cause of a seiche has disappeared, the oscillations gradually become damped, going through several cycles as they die away.

7.5 Movement in "Still" Waters: Under the Surface

The waves and seiches described in the preceding section happen at the surface of a lake. Other waves and seiches, internal ones, happen wholly within the lake, concealed beneath the surface. The notion seems extraordinary at first, until you consider what a surface really is: it is the interface between fluids of different densities. A lake's surface in the usual sense is the interface between the air above and the water below: both air and water are fluids. The contrast between their densities is extreme: water is nearly 800 times as dense as air.[24]

Layers of contrasting density are also found *within* a lake if the lake is stratified. At most levels, density increases gradually going downward, but the increase is comparatively abrupt at the thermocline, where warm, lightweight water floats on cooler, denser water; the change is abrupt enough to create a "surface." The surface is certainly not tissue-thin, as the surface between air and water is; nonetheless, it is thin enough to be shaped into waves and seiches known as *internal waves* and *internal seiches;* they are sometimes called *internal progressive waves* and *internal standing waves,* respectively.

Their behavior is very different from that of surface waves, because the density contrast between the upper and lower layers is, comparatively, so exceedingly small. This means that much less energy is required to create waves and that, once waves exist, the force tending to flatten them is only about one one-thousandth of the force tending to flatten surface waves. It follows that internal waves and seiches can become much bigger than surface ones. In some conditions[25] the energy that would cause 1-centimeter waves on the surface would cause 60-centimeter waves in the thermocline. Internal seiches 10 meters high are known to happen in Loch Ness.[26]

Internal waves are more sluggish, too: they are slower moving and have a longer period (the time it takes for one whole wave, from crest to crest, to pass a fixed point). For example, in Lake Mendota, internal seiches up to nearly 1.5 meters high have been recorded, with periods of 10 hours, making them ponderously slow compared with a surface seiche's 26-minute cycle on that lake. In a nutshell, what goes on underwater is more majestic than what goes on at the top—the water movements are of greater magnitude and happen at a more stately pace. It seems a pity that they are virtually invisible, but they do yield perceptible effects at the surface, as we shall see.

Consider internal seiches first; they are also called *temperature seiches*. As an internal seiche peaks at one end of a lake, the water surface vertically above it sinks slightly: that is, the internal seiche produces a very small "reflection" of itself at the surface of the lake.[27] What happens is shown in figure 7.7. In contrast to a surface seiche, in which water flows in only one direction at any one moment (see figure 7.6), it flows in opposite directions simultaneously in an internal seiche: while it is going from left to right below the interface, it goes from right to left above. The figure shows how this affects the surface level. The oscillations at the surface are minor, and sensitive instruments are needed to detect them.

Because of their greater amplitude, internal seiches are more important than surface seiches in their biological effects. Like overturns they redistribute dissolved oxygen and mineral nutrients. They also cause turbulence in the hypolimnion, preventing it from stagnating, at least temporarily.

Now consider internal progressive waves. Like internal seiches they are bigger and more sluggish than their surface counterparts. In a large body of water, they are sometimes five times as high, have periods ten times as long, and travel only one-twentieth as fast.[28] An example from the well-studied Lake Mendota, a small lake, describes some long, slow, small waves in the thermocline 11 meters down; their wavelength was 30 meters, their period 8 minutes, their forward speed 0.06 meters per second, and their height a mere 6 centimeters.[29]

Discovering the details is not a matter of simple observation, because everything is happening deep underwater. Subsurface waves are detected by noting how the temperature of the water varies over a period of time as measured

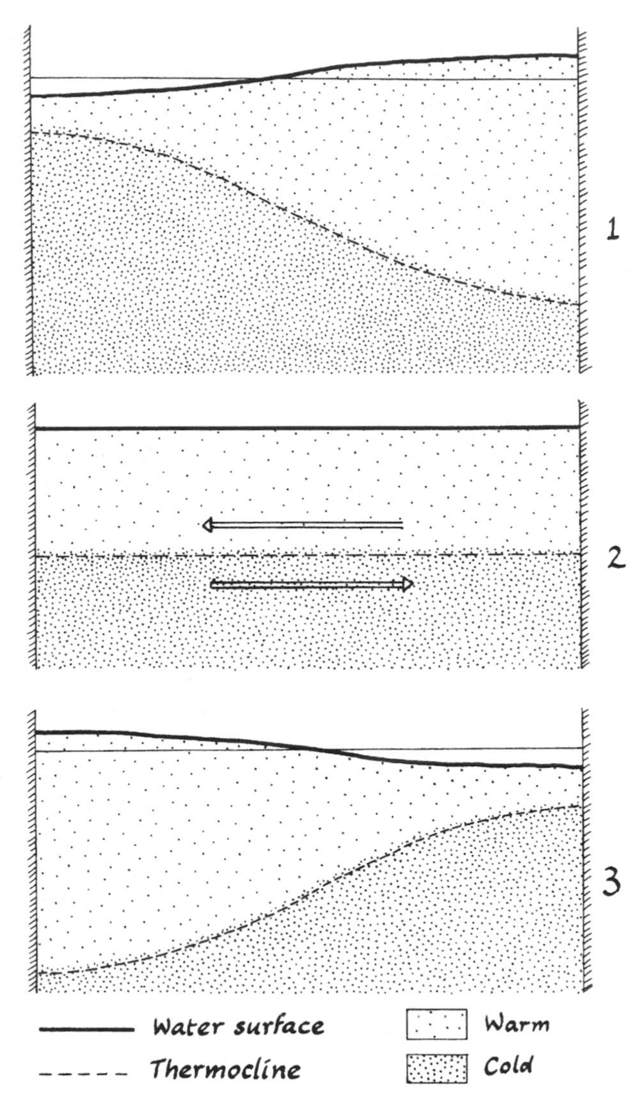

FIGURE 7.7. An internal seiche. The three diagrams show successive stages in one half-cycle. The seiche affects the thermocline; the slight ups and downs at the surface reflect the much larger movements below. Opposing currents flow above and below the thermocline. Their velocities are greatest when the thermocline is horizontal (arrows in middle panel); they slow to zero when the seiche reaches its maximum (top and bottom panels).

by a fixed thermometer several meters down. If waves are passing, the temperature goes up and down with extreme regularity; the thermometer records a high temperature while it is bathed in the warm water of the epilimnion, and a comparatively low temperature when the cool water of the hypolimnion rises to engulf it. The regularity of the ups and downs proves that underwater waves are responsible.

Internal waves are as varied as surface waves in shape. They can be smooth or choppy, and they can even build up into breakers. A breaker reveals itself when continuously recording thermometers at different depths show a sliver of cool water—the curling wave crest—momentarily above the warm water in the hollow of the wave.

While these events are happening invisibly, there is at least something for observers to see at the surface. Internal waves sometimes produce *slicks* (figure 7.8). These are long ribbons of smooth water winding across an otherwise slightly choppy lake surface. Provided the wind isn't strong, they form over

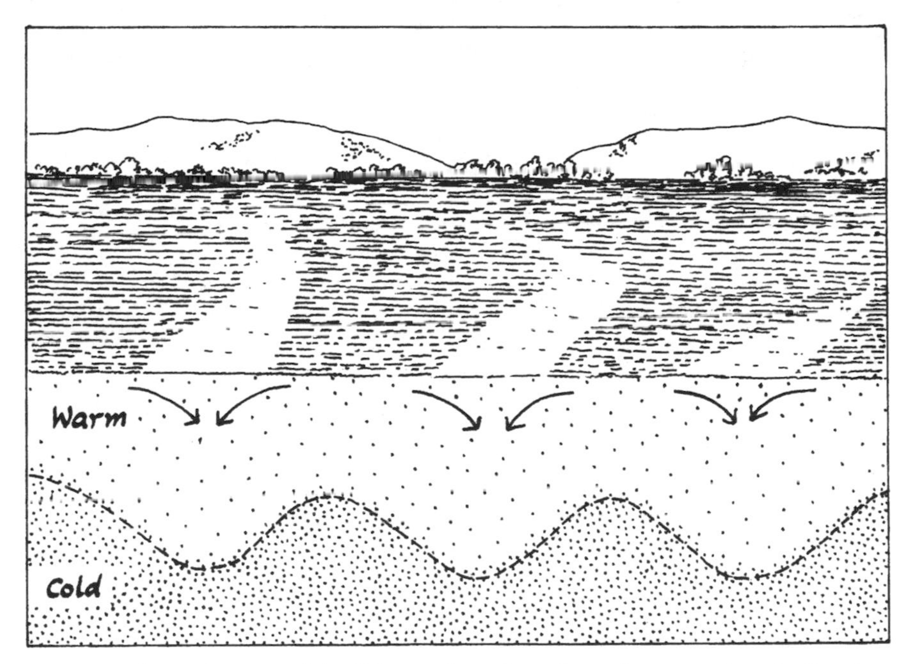

FIGURE 7.8. Internal waves and the slicks that form above them. The arrows show the direction of near-surface currents.

the troughs of internal waves where surface currents converge before flowing downward. If there is an oily surface film on the water, the film thickens where the currents meet, and is left at the top when the water sinks. In lakes used by power boats, the oil usually comes from the boats, but even pristine lakes often have a thin film of oil at the surface, produced by aquatic plants and animals. Whatever its source, an oil slick looks glassy.

Most slicks are glassy, but not all; flotsam of any kind can accumulate in slicks—dead leaves, fragments of water plants, even logs. Because they form above the troughs of internal waves,[30] a number of more or less parallel slicks are usually to be seen simultaneously, migrating slowly across the water in time with the slow-moving waves beneath. They drift in the same direction as the internal waves that cause them, which is not necessarily the same as the direction of the surface waves. The wind may have shifted since the internal waves were formed, in which case the surface waves, responding quickly to the wind, will no longer be parallel with the internal waves below.

Slicks can be several kilometers long and are usually spaced many meters apart. What could easily be mistaken for short, closely spaced slicks, directed across the waves, can often be seen on a lake, especially when the wind dies down in the evening and the surface wavelets become smooth; they show up best when the scene is lit by a setting sun; the smooth bands are not slicks, however, but *streaks* (also called *windrows*), and are formed in an entirely different manner. They owe their existence to the presence of *Langmuir cells.*

The downwind surface currents caused by wind drag often develop into helical currents, as shown in figure 7.9. The currents form the boundaries of individually circulating sausage-shaped "cells" of water, known as Langmuir cells, which lie horizontally, side by side, in a layer just below the surface. They are above the thermocline when there is one, and about 5 to 10 meters in diameter; their lengths are variable. Swarms of Langmuir cells often form, and once the wind has died down, they need not remain parallel.

The figure shows the direction of flow in each cell; the current flows partly across the wind and partly downwind (into the page). Where the currents in two adjacent cells converge and descend, as they do at every second cell contact, oil films and flotsam collect at the surface in the same way as they do in slicks. If an oil film collects, a short glassy "ribbon" appears on the water; this is a *streak* (it is unfortunate the word is so like *slick*). The name *windrow* seems preferable when the flotsam is more solid, or anyway three-dimensional—sometimes the flotsam consists of a froth of bubbles. The lines of foam that appear when a strong wind blows over water are windrows.

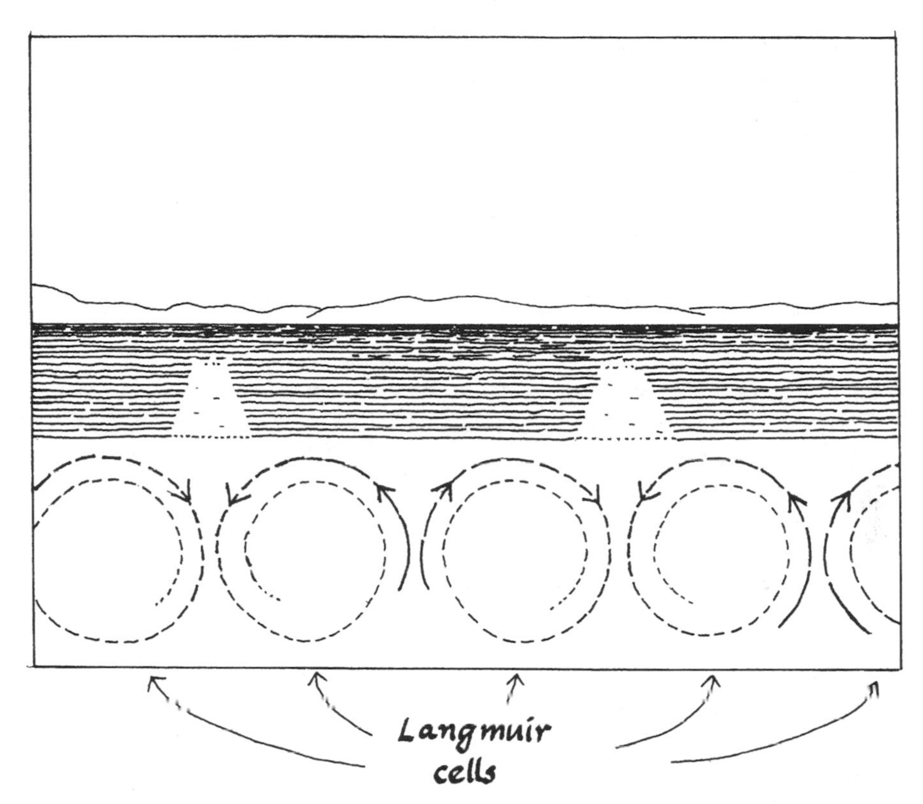

FIGURE 7.9. Langmuir cells just below a lake surface. The wind is from behind the viewer, and the helical currents around the cells are flowing forward into the page. Streaks at the surface form where the currents of adjacent cells converge; here they are oily and smooth. Note the different sizes of slicks and streaks. Slicks (figure 7.8) are usually five to ten times farther apart than streaks (this figure).

They are common at sea, where yacht sailors find them useful: experienced sailors know that currents paralleling the wind are strongest along the windrows, and weakest midway between them.[31] Kayakers and canoeists on lakes also use this knowledge.

7.6 Lake Water: What's in It?

However clear and pure the water in a lake may seem, it is never the same as distilled water. All lakes, without exception, contain other material as well.

It is just as well they do: without the necessary chemicals, aquatic life would be impossible.

Nowadays the innocent word *chemicals* has become a red rag of a word, associated with pollutants, pesticides, artificial fertilizers, unwelcome food additives, and toxic materials of all kinds. In truth, its meaning is far wider; it includes, among other things, the nutrients essential to all living things. In what follows, *chemicals* by itself is a neutral word. That said, let's consider where the chemicals in any lake come from, and what they are.

All lakes receive input from the atmosphere. The atmosphere everywhere contains particles of dust and particles from evaporated sea spray, even if only in minute amounts. The particles are washed out of the atmosphere by rain and deposited in lakes, either directly or by the streams flowing into them. The chemical elements arriving by this route include calcium, magnesium, potassium, sodium, and chlorine, the last two almost wholly from sea spray.[32]

Also from the atmosphere (and from other sources as well) come small amounts of some indispensable gases: oxygen, carbon dioxide, and nitrogen. Oxygen is only slightly soluble in water; nevertheless, when the wind ruffles a lake's surface, the water absorbs oxygen directly from the air; inflowing streams, especially swift, frothy, bubbling streams, are another source. Nitrogen, the gas forming nearly 80 percent of the atmosphere, is about one-half as soluble in water as oxygen is and enters lake water, as a dissolved gas, in exactly the same way. The fact is seldom mentioned in accounts of freshwater life, because nitrogen in its pure form (as distinct from nitrogen in various compounds) is useless except to bacteria; what they do with it is considered in chapter 11. Carbon dioxide, the major source of carbon for all green plants, dissolves in water three times as readily as oxygen, and although it forms only a minute fraction of the atmosphere (about 3 parts in 10,000), enough of it enters lake water to make a worthwhile contribution

So much for the "good" gases. Other gases, the pollutants that cause acid rain, come too: oxides of sulfur and nitrogen are the chief offenders. It's true that the "desirable" gas carbon dioxide forms a weak acid (carbonic acid) when it dissolves in water; it's also true that sulfur and nitrogen are elements necessary to life. All the same, their oxides, dissolved in rain, do immense damage to freshwater life. More on the subject in section 7.8.

The biggest outside source of lake chemicals, both good and bad, is incoming surface water, brought by rivers, streams, and overland flow. When water on its way to a lake flows over bedrock, or comes in contact with mineral grains in the soil, it dissolves numerous chemicals; many are only barely soluble, so that solutions of them are exceedingly dilute, but they are there nonetheless. For example, silica (quartz) and compounds derived from it are the commonest chemicals on earth but are among the least soluble; their concentration in surface waters averages only one part in 30 million.[33] Other "geological" chemicals are iron, sulfur, carbon (as carbonate and bicarbonate, chiefly from limestone and dolomite rocks), and the elements listed previously as present in atmospheric dust; the dust—at least the "natural" part of it—comes from surface rocks.

In addition to inorganic chemicals derived from minerals, the surface water entering any lake will have picked up organic chemicals from the soil; these are produced by the decay of dead plants and animals. They are natural chemicals, needed as nutrients by the lake's aquatic ecosystem, and they are therefore desirable, so long as they are not present in excess. Lakes and ponds in the cold regions are often fed by streams that have flowed through peat, from which they have picked up dissolved organic carbon—*DOC* as it is called[34]—which often turns the water brown like tea. So much for natural chemicals in surface water. Unnatural chemicals, in other words, pollutants, are also brought in, as we shall see in section 7.8.

Incoming groundwater is another supplier of dissolved chemicals. The inorganic chemicals in groundwater are of the same kinds as those in surface streams, but often they are at higher concentrations. This is particularly true if the water has traveled underground slowly, for a long distance; it will have had time to pick up comparatively large amounts of dissolved minerals.

Some of the chemicals in lakes come from lake-floor sediments; many of them came, originally, from living or dead organisms and remain absorbed by the mud on the bottom of the lake for a time. Such chemicals are sometimes released from the mud again, and reenter the water.

This brings us to life and death as the source of the most important, from a biological point of view, of a lake's chemical content. Growing organisms must obviously get their body-building materials from somewhere; in other words, they must have certain elements to live. When they are dead, their chemicals are recirculated.

The chemicals needed in comparatively large amounts by living things are hydrogen, oxygen, carbon, nitrogen, and phosphorus. Hydrogen is hardly ever mentioned as one of the requirements of aquatic life, although it is one of the chief ingredients in any living body, animal or plant. Its presence is taken for granted because it is one of the two elements that combine to form water (oxygen is the other); as hydrogen is always absorbed and used as water, there's never a shortage (unless the lake dries up!).

Oxygen is a different matter. The oxygen atom that forms part of every water molecule is inaccessible to living organisms, which must have "free," uncombined oxygen.[35] The needed free oxygen is available in some parts of a lake, most of the time; some of it has dissolved into the lake water directly from the atmosphere, as we saw above, and some of it is in the midst of being recycled—it is being given off by green plants as a by-product of photosynthesis, the process by which plants extract carbon from carbon dioxide.

Carbon, the third of the elements required in bulk, comes from carbon dioxide dissolved in the water; part of it comes from the atmosphere and part from living organisms in the lake. The vast majority of organisms (including green plants, when it's dark) absorb oxygen and give off carbon dioxide in the act of breathing. Only green plants give off more oxygen than they take in, and then only in daylight.

It's even possible to set up a sealed aquarium that requires no oxygen or carbon dioxide from the outside world; green plants and nongreen organisms maintain themselves in it indefinitely, by continually exchanging oxygen and carbon dioxide in their own self-contained world.

Finally we come to nitrogen and phosphorus; they deserve special consideration from the biological point of view, because one or the other is often a lake's limiting element. This means that the quantity available is what sets an upper limit to the volume of living matter that a lake can support. A gardener will notice that they are the chief elements in the commonest garden fertilizers: a well-watered garden never lacks for hydrogen, oxygen, or carbon, but it is bound to be unproductive if nitrogen or phosphorus are in short supply. The same is true of lakes. Lakes differ from one another enormously in their nitrogen and phosphorus endowments, and these differences account for much of the wide variability in lake ecosystems.

Nitrogen is naturally present in a lake in various forms: as pure nitrogen, dissolved from the atmosphere; in inorganic nitrogen compounds dissolved from the rocks; and in organic nitrogen compounds in the excreta of all aquatic animals and the decaying remains of all aquatic life. The great majority of plants can use only inorganic compounds as a source of the nitrogen they must have as a constituent of their proteins. The other kinds of nitrogen have to be worked over by bacteria of various kinds before they become available for animals, as we shall see in chapter 11.

Phosphorus is, in a sense, the most important (biologically) of all the elements found naturally in a lake, because it is the one that produces the most dramatic effects. Too little phosphorus can make an otherwise well-endowed lake unproductive—that is, phosphorus is limiting. Too much of it, often the result of pollution, can disrupt a lake's ecosystem. What happens when naturally unproductive lakes are fertilized with phosphorus has been discovered by experiments carried out on whole, natural lakes.[36] The lakes used were ice-scour lakes on the Canadian Shield in northern Ontario—clear, unpolluted lakes in basins of hard rock. A lake experimentally polluted with phosphorus promptly developed a spectacular bloom of blue-green bacteria; it contrasted conspicuously with untreated lakes in the neighborhood.

Lakes vary tremendously in their productivity, or in other words, in their ability to support life. As a general rule lakes near the headwaters of a watershed, especially in mountainous country, are the least productive, and lakes at successively lower levels in the same watershed become more and more productive.[37] This is what you would expect. The water entering a mountain lake is almost pure: it comes from rain and snow and from streams flowing over clean rock. A lake at a lower elevation contains numerous impurities. It is fed by streams and overland flow that have picked up organic and inorganic compounds from the soil and the vegetation; sewage and chemically enriched groundwater may also have seeped into it. The result is a much "richer" lake, well supplied with the nutrients that make it productive.

Figure 7.10 shows the typical appearance of a productive and an unproductive lake. The contrast is familiar to everyone who enjoys hiking, boating, or fishing. In technical terms, a productive lake is described as *eutrophic* and an unproductive one as *oligotrophic*. Lakes between the extremes are *mesotrophic*. Although one usually thinks of oligotrophic lakes as exceptionally clean and clear, this isn't always so. Perhaps the least productive, most oligotrophic, of all lakes are those within a few meters of the end of a glacier; their waters are an opaque milky-green because they are full of the exceedingly fine-grained rock flour created by the grinding of the glacier on bedrock.

Oligotrophic and eutrophic lakes differ from each other in several ways, which we consider in the next section.

7.7 The Ever-Changing Architecture of Lake Water

The various materials dissolved in lake water, mentioned in section 7.6—oxygen, carbon dioxide, and all the rest—don't just sit there. Many of them are as active as the living organisms that depend on them for existence, continually engaging in chemical reactions, increasing and decreasing in concentration, or shifting their locations within a lake.

We considered earlier (section 7.3) how lake water is stratified: because

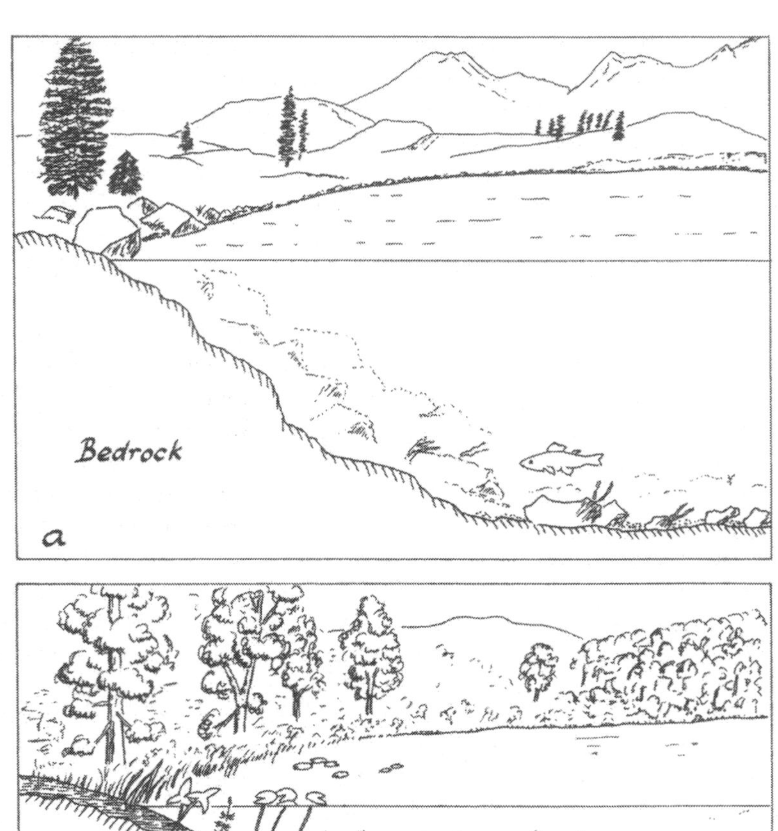

FIGURE 7.10. *(a)* An oligotrophic lake. *(b)* A eutrophic lake.

warmer water floats on cooler water (provided all the temperatures are above 4°C), the temperature in a lake decreases from the surface down to the floor. But it isn't only temperature that is stratified: the concentration of dissolved oxygen is stratified too. This is because oxygen dissolves more readily in cold water than in warm; if a given volume of water is cooled from 20°C to 5°C, the quantity of oxygen the water can hold in solution increases by more than 40 percent.[38]

Besides temperature and oxygen concentration, another factor that changes with depth—one that is easy to observe—is the light intensity. Brightness decreases from the surface downward in a way that varies enormously from one lake to another. As any swimmer knows, in clean, clear water, sunlight penetrates deep, and you can see a long way down. But if a lake is stained or turbid, the light cannot penetrate far, and darkness prevails except in a shallow layer at the surface.

Now recall how markedly lakes differ from one another in the concentrations they contain of the essential nutrients nitrogen and phosphorus; the quantities available determine the productivity of a lake. Productivity depends, as well, on light, oxygen, and warmth. Green plants including microscopic algae (a mixed collection of photosynthetic organisms) cannot live without light. Animals—everything from fish to amoebas—cannot live without oxygen, and all the physiological processes that happen in living organisms proceed at a pace controlled by temperature. It follows that in a stratified lake the living conditions are also stratified; indeed a lake, regarded as the home of a highly diverse community of organisms, can be thought of as a structure with a definite architecture.

One of the most important "architectural features" of a stratified lake is the thermocline (see figure 7.4). This marks the separation between warm water above and cool water below. Another, equally important feature is the *compensation level.* This is the level above which the light is bright enough for photosynthesis to take place.[39] In most lakes the compensation level is above the thermocline,[40] but markedly oligotrophic lakes are an exception—their clear waters usually allow light to penetrate below the thermocline in amounts useful to plants.

A third, more diffuse level is that at which the concentration of dissolved oxygen is greatest: the concentration increases from the surface downward because of the falling temperature, but not all the way to the bottom; below a certain rather ill-defined depth, oxygen concentration decreases, regardless of the temperature, for two reasons. First, the deeper water is less often in contact with the atmosphere, and so less able to absorb atmospheric oxygen. Second, as you go deeper, and the light steadily decreases, the rate at which oxygen is produced by photosynthesizing green plants falls more and more behind the rate at which it is consumed by nongreen organisms. Therefore, the greatest oxygen concentration lies somewhere between the surface and the bottom.

The living organisms in a lake and the three stratified factors we have just considered—temperature, light, and dissolved oxygen—all influence one another. Long chains of cause and effect often operate, all ultimately controlled by the productivity of the lake, that is, by the quantity of nutrients it contains. "Rich" (eutrophic) and "poor" (oligotrophic) lakes differ from each other in many ways depending on how the nutrients and the stratified factors interact: lakes at opposite ends of the rich-poor spectrum differ radically in their architecture. Let us look at the extremes.

First consider a eutrophic lake, richly supplied with nutrients. Many big-leaved water plants grow in the shallow parts of the lake in summer; some familiar examples are water lilies, water shield, and pondweeds. Their leaves either float on the surface or a short distance below. When the weather is sunny and warm, the leaves carry on photosynthesis at a rapid rate, liberating quantities of oxygen (only a fraction of this oxygen dissolves because the water is so warm; the rest collects as bubbles around the leaves' margins). The leaves shade the water below them, however, bringing the compensation level much higher than it would be if the leaves were not there, and this means that photosynthesis becomes impossible only a short distance below the surface. At the same time, quantities of detritus—fragments of dead leaves, the feces of all the tiny invertebrate animals living on the undersides of floating leaves, the bodies of drowned insects, and so on—form an organic "rain" that sinks slowly downward through the shaded waters, being consumed by bacteria and microscopically small fungi as they sink. The bacteria and fungi, collectively called *decomposers,* are living, breathing organisms that continuously use up the oxygen in the water. This causes the oxygen concentration to fall off rapidly at increasing depths.

In the deeper parts of the lake,[41] parts too deep for rooted plants to grow, leaves are absent and their place is taken by a cloud of microscopically small organisms, some of which carry on photosynthesis, while the others feed on the photosynthesizers (taken all together, these organisms constitute the lake *plankton,* about which more in chapter 11). Plankton casts negligible shade, so the compensation level is comparatively deep; but the plankton does generate a rain of small bodies, fodder for the decomposers below, so that the oxygen concentration again falls off rapidly.

Now consider a nutrient-poor, oligotrophic lake. The architecture produced by stratification is entirely different. Living things are scarce in such a lake. If the water is clear, sunlight penetrates deep, warming the water to a considerable depth; that is, the thermocline is far down. The scarcity of photosynthesizing greenery (ordinary plants and the "green" part of the plankton) means that little oxygen is generated in the water, but this is more than made up for by the absence of a rain of detritus and the decomposers feeding on it; as a result the dissolved oxygen is not used up, and the oxygen concentration remains relatively high for a long way down. In a cool climate, where the water remains cold despite the deeply penetrating sunlight, the conditions are ideal for trout.

We now come to an "architectural feature" we have not considered yet: the mud on the floor of the lake, known to ecologists as the *benthic zone.* The ingredients of the mud vary from lake to lake. In clear, cold lakes the chief ingredient is usually inorganic silt and clay brought in by rivers and streams; not much lives in such sediments. In warm, productive lakes organic sediments predominate. The lake bottom is where the rain of organic detritus typical of eutrophic lakes comes to rest, and where most of the decomposition goes on. It is performed by myriads of living, oxygen-breathing microbes, chiefly bacteria, that live in the mud and consume the detritus.

If the detritus rain keeps on falling for a long period, as happens when the weather stays warm and calm, with no wind-caused currents overturning the stratification, the oxygen in the lowermost water layer becomes exhausted. Then most of the bottom-dwelling aquatic animals suffocate, as do the oxygen-requiring bacteria. The only survivors are bacteria that can live without oxygen or that obtain their oxygen from other sources (see chapter 11).

Some of the bacteria in lake-bottom mud play a crucial role in cycling phosphorus; they extract "used" phosphorus from organic detritus and reconstitute it as a phosphate usable by plants, including the minute green algae of the plankton. If the bottom water has become deoxygenated, the phosphate dissolves in the water, whence the plankton absorbs it. But the phosphate does not dissolve readily if the water in contact with it is still oxygenated; it then remains trapped, unavailable to the plankton photosynthesizing in the sunlit water far above.[42]

This is an example of the complicated chains of cause and effect that sort lake water into layers differing from each other chemically and physically. In this case too much oxygen at the bottom prevents phosphate from reaching the surface, where it is needed by the "greenery" that produced the oxygen in the first place. The deadlock is finally overcome when something—a strong wind perhaps—causes the lake to overturn; this disturbs the mud, breaks the stratification, and mixes all the water from top to bottom.

Up to this point we have considered lakes that overturn completely at least once a year; in shallow, open lakes, exposed to the wind, overturns sometimes happen every few days or even more frequently. By contrast, in lakes that are unusually deep and sheltered, the deepest layers don't participate in overturns: they remain stagnant for years, while the upper layers behave in the ordinary way.

Prolonged—even permanent—stratification also happens if a lake is *chemically stratified,* that is, if the layer of water at the bottom is dense because of dissolved chemicals. The chemicals sometimes enter in groundwater seeping up through the lake floor, and sometimes they are the accumulated products of decomposing detritus that have become so concentrated that they affect water density. Whichever the cause, the result is the formation of a *chemocline,* separating relatively pure water above from water with dissolved chemicals below.

Lakes having a permanent, dense, stagnant layer, devoid of oxygen, at the bottom are known as a *meromictic.*[43] And if the dense layer owes its density to dissolved chemicals, the lake has both a thermocline and a chemocline. The latter is the deeper "cline," and only below it does the water remain stagnant for years. The water above it becomes stratified in the ordinary way and overturns seasonally. In a sense, meromictic lakes can be thought of as two separate, independent bodies of water, one atop the other. The upper water body is an "ordinary lake," which floats upon, but never mixes with, the lower body, a dark, dense, cold, deoxygenated, stagnant, unseen "pool;" whose existence far below the sparkling, sunlit surface of the water is unknown to most observers.

7.8 Pollution

This chapter would obviously be incomplete without a section on pollution. As the human population grows and spreads, more and more of the world's natural water is put at risk. We considered groundwater pollution in chapter 3; here we take up the topic of surface-water pollution, especially of lakes but also of rivers from which many lake pollutants come.

Nowadays so much is said and written about pollution by the gullible for the gullible that it pays to be a cynic, or at least to attend carefully to what is said and by whom. If an audience is told that a certain lake contains 0.001 parts per million of arsenic, let's say, some gasps of outrage will probably be heard, even though a concentration twenty times as great would be harmless according to most authorities. Figures are meaningless until they are intelligently weighed by someone knowledgeable enough to measure the risks dispassionately.[44] That said, let us proceed.

The routes by which pollutants enter a lake are the same as the routes by which the water enters; they are shown in figure 7.3. The sources are the same as the sources of groundwater pollution: they are listed in section 3.5. The pollutants themselves are legion. A convenient way to subdivide them is to consider, first, materials that only become pollutants when present to excess—in moderate amounts they are desirable. After that, we come to materials that are unwanted at any concentration.

The nutrients that nourish a productive aquatic ecosystem are all well and good in "natural" quantities. In many lakes, nowadays, they are present to excess, and the result is "galloping eutrophication,"[45] otherwise called *hypertrophication.* The causes are, first, sewage (plus runoff from feedlots) and, second, agricultural fertilizer washed in from nearby fields (plus runoff from golf courses, ornamental parks, and gardens). These nutrients enrich any body of water they reach with lavish amounts of nitrates and phosphates. Sewage contains a much higher proportion of phosphates than does natural water, in which algal growth is usually limited by phosphorus shortage. Dense blooms of algae therefore grow in lakes polluted with sewage. Who has not seen the result—a surface covered with warm, green scum, lumpy with trapped bubbles? As the algae die and decompose, the oxygen in the water is used up, and the lake's fish, tadpoles, and invertebrate animals suffocate. The rate at which oxygen is used up by decomposers, called the *biochemical oxygen demand* (BOD), is used as a measure of this kind of pollution.[46]

Any algal bloom dies down in time, and certainly disappears when the weather turns cold. But the surplus phosphorus that caused it has not gone away: it remains in the mud at the bottom, ready to be recycled.

In some ways, treated sewage is as serious a pollutant as raw sewage. It is true that fecal coliform bacteria—the best known being the notorious *Escherichia coli*—are found only in raw sewage, which lulls people into believing that treated sewage is harmless. Not so: treated sewage is as rich as the raw variety in unwanted nutrients; moreover, in treated sewage they have already been converted by bacteria into a form immediately usable by plants.[47]

The lack of oxygen that results from overnourishment of a lake often yields unwanted by-products. In the absence of oxygen, bacteria that produce methane and hydrogen sulfide go to work. Methane is a powerful "greenhouse" gas that is believed to contribute appreciably to global warming.

Next we come to pollutants that are harmful without being toxic. One is the unnaturally warm water that has circulated as a coolant in thermal generating stations or other industrial plants, and then been discharged into a natural body of water. The excess warmth causes the receiving water to lose some of its dissolved oxygen (because oxygen is more soluble in cold water than in warm) and become uninhabitable by many of the creatures that normally live in it; in a nutshell, it destroys the natural ecosystem.

Another pollutant that isn't toxic in the ordinary sense is excess sediment entering lakes and streams, a consequence of clearcut logging, road building and other construction work near a shoreline, and natural and unnatural landslides. Sudden, abnormal influxes of sediments can destroy an aquatic ecosystem within minutes by blanketing the bed of a lake or stream and suffocating all the bottom organisms. Landslides often poison as well as suffocate: they are apt to contain pollutants such as oil, gasoline, paint, and solvents.

This brings us to toxic pollutants. Their number is vast. Who has not heard of PCBs, dioxins, furans, mercury, lead, and scores of others? The most serious toxic pollutants in the Great Lakes[48] have reached the lakes from industrial effluents discharged directly into the lakes, from polluted groundwater, and from polluted overland flow; they are also carried in from a distance by tributary streams and rivers, and by the wind, as atmospheric pollutants. Lake Erie is the worst of the lakes; it is probably the world's biggest settling pond.

Attempting to remedy the deplorable situation in the Great Lakes has become an industry itself. It is not enough to ban the discharge of persistent toxic materials from this time forth: huge amounts are already present, trapped in the lakes' sediments, including materials such as DDT and PCBs that have long been banned; they are difficult or impossible to eliminate. To dredge them up and dump the dredge spoils in leaky landfills is no help; it merely recycles the pollutants. The danger they present is that, even if they're harmless while they're buried, the sediments can become resuspended in the overlying lake water at any time, for any of a number of reasons. Big storms, the passage of ships, and dredging all stir up lake-floor sediment; so does *bioturbation,* the constant disturbance of the mud by burrowing aquatic animals.

The enormous difficulty—perhaps the impossibility—of remedying past mistakes makes Great Lakes pollution much more serious than is generally realized. This is probably true, also, of numerous smaller, less well-studied lakes. Many of the toxic chemicals now known to be in the lakes are acutely poisonous, even at low concentrations. Others may be present at still lower concentrations, so dilute as to be undetectable. This does not mean they won't gradually become concentrated, by the process of *biomagnification.* Many poisons are ingested, with their food, by the minute plankton organisms at the bottom of the aquatic food chain, and become incorporated in the animals' tissues rather than being excreted. The poisons are then passed from level to level up the food chain, from plankton to small fish to large fish to birds, becoming more concentrated at each step: the total increase in concentration is sometimes more than a millionfold. It has been said that "a person who eats . . . lake trout from Lake Michigan will be exposed to more PCBs in one meal than in a lifetime of drinking water from the lake."[49] For a fish or bird with no alternative source of "meals," the results can be catastrophic: published photos of grossly malformed animals, for example cormorants with crossed mandibles, are widely familiar.

Persistent poisons, continuously accumulating, are a lasting threat to our freshwater supplies. They may persist for decades or centuries after being deposited. Poisons accumulate in a lake's water as well as in its sediments, especially in deep lakes with long residence times. They accumulate, too, in the deeps of meromictic lakes, whence they could become liberated by an unforeseen accident, say a lowering of the lake's level by a natural or artificial disturbance.

A form of pollution that scientists predicted for decades before the general public finally got the message is *acid rain.* Its existence illustrates two important facts about pollution. First, that water pollutants come in very significant quantities from the air, itself polluted by industrial smokestack emissions and vehicle exhausts. Indeed, the majority of pollutants are airborne to some extent; airborne contributions add appreciably to the quantities arriving by other routes. Acid rain is made acid when certain industrial gases polluting the atmosphere, chiefly sulfur and nitrogen oxides, become dissolved in falling rain; when this happens, the rain becomes acid,[50] and so does the surface water it finally reaches, with disastrous consequences for aquatic ecosystems.

The second point to notice is that sulfur and nitrogen are among the mineral nutrients essential for life; so is oxygen (though some kinds of bacteria cannot tolerate it). Even so, certain compounds constructed of these elements, those in acid rain for example, are extremely damaging. Polluting metals, too, are usually much more toxic in some compounds than in others. One of the harmful effects of acid rain is that it often reacts with mildly toxic metal compounds that happen to occur naturally in a lake, converting them to more toxic forms.[51]

Acid rain is not the only cause of unnatural acidification in surface waters. In some areas *acid mine drainage* is the cause. Many coal deposits, and many mineral ores, contain sulfides: iron sulfide—otherwise, pyrite or "fool's gold"—the mineral used for producing sulfuric acid commercially, is a familiar example. Sulfides are often abundant in mine wastes, the mounds of broken rock cast aside during mining operations, and their first exposure to the air and its oxygen comes when they are dug from the ground and broken into fragments. The sulfide combines with oxygen and water, yielding sulfuric acid which drains into nearby streams and lakes. The reaction is greatly speeded up if bacteria of the appropriate species are present (see chapter 11). Acid rain pollutes lakes in remote wilderness areas, and acid mine drainage pollutes those in areas that would be wilderness but for the presence of an isolated mine.

Another type of pollution in unpopulated areas is mercury pollution resulting from the creation of reservoirs; but reservoirs are not true lakes, and I postpone a consideration of them to chapter 9.

8

When Water Freezes

8.1 How Water Freezes

In the "north country," as all North Americans well know, a time comes in the fall when the setting sun is no longer reflected as a glitter path across every lake, and the murmur of creeks and streams falls silent. The sights and sounds of liquid water will be absent until the following spring. It is time to contemplate water in its solid rather than its liquid form.

The way ice forms on small, still ponds is familiar to everybody. The first ice is a paper-thin layer, so clear as to be almost invisible. It is often called "black ice" because its transparency allows you to look right through it into the darkness of the water below.

Few people pause to ask why the first ice on still water should always be at the surface. The answer is that the water of a calm pond is stratified, as shown in figure 7.4. Once the temperature of a cooling body of water is nowhere higher than 4°C, the coldest water—the water that will freeze first—is also the least dense and therefore collects at the surface.

A clear "window pane" of ice cannot form on water stirred by the wind, however. Wind action overturns the water, mixing it and making the temperature the same throughout; all the water reaches freezing point simultaneously, and there is no extracold surface layer. The constant movement prevents the mass of cold water from freezing solid. Instead, scattered ice crystals form and move around independently. This is *frazil ice*. The ice crystals float to the surface because ice is always lighter than water, and there they accumulate as patches of frazil slush, which grow until they coalesce and congeal. The result is a layer of opaque, gray ice on the pond surface: opaque because the light is reflected by a myriad separate ice crystals, gray because only a fraction of the daylight is reflected while the remainder penetrates to the water below.

8.2 Fall Freeze-up

In the north, fall freeze-up affects all fresh water. Lakes freeze later than ponds, and flowing water later than still water, but where the winter is cold enough, ponds, streams, lakes, and rivers are all involved eventually. In this section we consider the course of events on ponds and lakes first, and then in flowing water.

Once fall freeze-up has begun, the first-formed ice on a lake spreads and grows. It doesn't take long for a lake of moderate size to become totally ice-covered, after which the ice thickens. New ice is added to the bottom of the old, sometimes in the form of frazil slush, or sometimes—if the water is still—as crystalline columns of ice.[1] As the ice thickens, it turns from gray to white because more and more of the sunlight falling on it is reflected; less and less gets through to the water below, and the underwater world dims and darkens.

New ice is also added on top of the original layer. Some of it arrives as snow, which becomes compacted into ice. And some is "overflow ice," formed if the load of snow is heavy enough to break the existing ice; then water seeps up through the cracks, floods the surface, and freezes. The snow on a lake is often unevenly spread; it may be swept off the center where the wind is strongest, and deposited in drifts near the shores where the wind slows down. If the original ice was black ice, the frozen lake surface may be dark at the center and white around the shores. Ice thickness is usually less on a snow-covered lake than a bare one; this is partly because light, fluffy, newly fallen snow insulates the ice and slows the rate at which new ice is added at the bottom, and partly because the snow reflects the sun's rays instead of absorbing them.

The thickness of the ice at winter's end depends on how long, and how cold, the winter has been. Where winters are not interrupted by mild periods, freeze-up lasts until spring, and while the cold continues, the ice never stops growing. The ice on arctic lakes often thickens to 2.5 meters or more.

Although a frozen lake looks utterly still, hidden movements may be going on below the ice-covered surface. Slow currents may be flowing.[2] The temperature of the water is just above the freezing point directly below the ice, and it increases downward to 4°C at most (see section 7.3). If the solid floor of the lake is at this temperature or lower, nothing happens (or anyway, no movement: the water will continue to cool if it is warmer than the lake bed).

But if something warms the bottom water, currents begin to flow. The "something" may be stored warmth in the lake sediment, left over from summer; it may be the earth's internal heat; or it may be the sun's warmth penetrating the ice, which can happen just before spring breakup. Whatever the cause, if the water under the ice is warmed where it comes in contact with the lake bed, it will begin to move: the way it moves depends on the temperature of the bottommost layer of water. If this water is at 4°C to begin with, the warming will make it less dense and cause convection currents to rise. But if it is cooler than 4°C, and the warming raises its temperature to (but not above) that critical level, it will become denser than it was, causing it to flow along the bottom in whatever happens to be the downhill direction.[3]

Under-ice currents are detectable only with the appropriate instruments; the human observer sees nothing, but fish and other animals are affected. Most frozen lakes seem to be motionless and quiescent through the months of winter, but this is not true of mountain lakes drained by steep, swift streams. As the level of the water supporting the ice on such a lake slowly drops, the ice sags in the middle; from being perfectly flat, as an ice sheet "ought" to be, it takes on the shape of a shallow bowl, a surprising sight until you realize the cause. Sagging ice is especially common at high elevations, where winters are cold and the stream draining a lake may be too steep and rapid to freeze. The sagging ice scrapes the lake floor in the shallows near the shore and prevents water plants from growing.[4]

Rivers and streams freeze later than lakes in the same neighborhood: the faster the current, the more delayed the freeze-up. When winter sets in, flowing water first becomes supercooled; that is, it remains liquid because of turbulence even when its temperature falls below the freezing point.[5] It cannot supercool indefinitely, however; when it is sufficiently cold, crystals of frazil ice form and stick to whatever they touch.[6] Some stick to each other, forming patches of floating slush that grow into *ice pans* (figure 8.1). The pans brush against each other as they float downstream, knocking off projections until all are smoothly rounded. Some freeze to the river's banks. Because of the mixing, there is no comparatively warm water at depth: rather, the water is at freezing point or below all the way to the bottom, chilling the riverbed;

FIGURE 8.1. *(a)* Freeze-up on a river in late fall; the first ice takes the form of thin ice pans, rounded by repeated collisions. *(b)* The same river after spring breakup. Thick ice floes float down the river; some were deposited on shore, when the river level was high owing to a temporary ice jam downstream.

then, if the bed becomes cold enough, frazil crystals and slush freeze directly onto it as *anchor ice,* which covers the bottom with a slippery white coating.

Meanwhile, at the surface the ice forming a border along each bank grows toward the river's center. The width of the unfrozen water surface shrinks, and more and more of the ice pans floating along on the current become hung up; the various detached pieces of surface ice stick to each other and grow together, until finally the ice cover is complete. The surface ice then goes on thickening in the same way that lake ice does: frazil ice in the water adds itself to the underside, while snow settles on top and becomes compacted. As with lakes, the weight of the snow sometimes cracks the ice it is lying on, allowing the water below to flow onto the surface and freeze.

A frozen river is also apt to overflow upstream of spots where thickening ice has narrowed the channel. In shallow stretches the river may freeze right to the bottom, damming it completely; then all the water still flowing overflows onto the floodplain, where it freezes to form a sheet of ice known as *aufeis,* or as an *icing.* Overflows often happen at the same place several times in a winter, building up layer upon layer of aufeis; in the Arctic, expanses of aufeis often become thick enough to last well into summer, or even into the following winter.

Unlike a lake, whose freeze-up date is determined by local temperatures, the freeze-up date of a stretch of river depends on the history of the water—on where it has come from and how long it has taken to come.[7] Many rivers in arctic Canada and Alaska flow northward. They flow from warmer latitudes and carry their warmth into the north with them, postponing freeze-up. This effect is offset, however, if a river receives cold tributaries. This happens in the Mackenzie River, for example, which flows north from Great Slave Lake; it would retain its summer warmth much longer than it does if it were not cooled by cold tributaries coming from the high mountains to the west.

8.3 Spring Breakup

For northerners, breakup is the first sign of spring. In this section, as in the preceding one, we consider lakes first, then rivers.

The first part of a frozen lake to melt when the weather warms is the snow layer on top of the ice.[8] Throughout the winter the snow has been warmer at the bottom, at its contact with the ice, than at the top, at its contact with the air; but when the air warms in spring, this gradient becomes reversed—

the temperature of the snow becomes greater at the top than at the bottom. The change resembles overturn in temperature-stratified lake water (see figure 7.4), though in the case of snow nothing moves.

On clear days the sun's rays warm the snow directly, but it melts or evaporates slowly as long as the air is still cold and while the pure white surface is reflecting the sun's rays back to the sky without absorbing any warmth. As the days lengthen and the sun's midday elevation increases, however, the air itself warms up, and snow on the land begins to disappear. On gray days the warmth radiated by dark, snow-free land is reflected downward by the clouds; it warms the frozen lake, and what's left of the snow on land. Precipitation now comes as rain, which further hastens the melt.

A stage comes when all the snow is gone, from the land surrounding the lake as well as from the lake ice. Now melting of the ice begins in earnest, starting at the lake's shores which, being dark, absorb the sun's rays and warm quickly. This is why a moat of water surrounding the ice is the first sign that breakup is in progress. The ice, now floating free, is no longer brilliantly white and no longer reflects nearly all the sun's visible and short-wave (ultraviolet) radiation.[9] Some of the ultraviolet penetrates the ice, melting it internally and warming the water underneath.

Melting now continues in the interior of the ice and at its upper and lower surfaces. Researchers in the Canadian Arctic Islands measured the relative speeds of melting at the three locations, and found that 50 percent of the melt was internal, 40 percent at the upper surface, and 10 percent at the lower.[10] These findings don't necessarily apply in other latitudes.

The quality of the ice changes as it melts. While it is frozen hard, much of it consists of long, upright, hexagonal crystals. Melting begins at the crystals' vertical surfaces, where they touch, separating them from each other and turning them into so-called *candle ice;* meltwater from the surface trickles down between the fragile "candles," which are often no thicker than kitchen matches. When the water in the moat around the lake ice island becomes choppy, the candles collapse, jingling and tinkling as they fall; big segments of the lake's ice cover, made up of innumerable candles, don't take long to disintegrate and disappear.

Candle ice is almost as transparent to sunlight as water is, and in a large, deep lake, this occasionally allows the water below the ice to acquire its summertime stratification (warm water above, cooler below) before the ice disappears.[11] When this happens, there is no spring overturn: when the ice finally melts, exposing the water to the wind, the stratification is already too strong—the density gradient too pronounced—for the wind to do more than merely churn its warm surface layer; and when overturn fails, reoxygenation of the water fails too. Early stratification such as this, ahead of spring breakup, is unusual, however. In the majority of lakes, spring overturn plus oxygenation happens after breakup, and summer stratification develops after that. Moreover, in shallow lakes reoxygenation can begin before breakup, because green plants and other photosynthetic organisms in the water start to photosynthesize, and produce oxygen, by the light of the spring sun shining through the clear ice.

River breakup happens in two quite different ways: the ice may simply melt, or it may be broken to pieces and carried off by the current. In other words, there are two kinds of breakup, *thermal* and *mechanical;*[12] a given stretch of river may break up thermally in some years and mechanically in others, depending on the volume of the meltwater and the rate at which spring temperatures rise.

Thermal breakup is not unlike breakup on a lake. It happens in gently flowing rivers, particularly in years when the load of snow on the ice has not been heavy enough to cause cracks. Most of the ice melts in place, without shifting. Only when it is thin and weak does it disintegrate and float away.

Mechanical breakup is a far more dramatic affair. It is the kind of breakup usual in powerful, fast-flowing rivers. What typically happens is this. The thick midwinter ice sagged and cracked when the water level was low during the coldest months, leaving the ice unsupported; the cracks tend to be near the banks and parallel with them. When the rising waters of spring lift the ice, the cracks open: the central part of the ice floats up because of its buoyancy, while strips of ice at the margins stay put because they are frozen fast to the banks. Water floods the marginal strips, forming *shore leads.*[13] The central ice is now free to float with the current. The rising water level widens the river, giving the floating ice sheet room to move. If the current is powerful—and it often is, because of the abundant meltwater—it breaks the ice, first into big plates, and then into numerous *ice floes.*

Conditions now exist for an *ice jam* to build. Floating floes become hung up in shallows, or as they round sharp bends in the river, creating obstacles on which succeeding floes pile up. Or floes from an open stretch of river may be driven against a still solid stretch of ice, pile up on it, and smash it, producing a heap of solidly interlocking floes; the same thing can happen where a thawed tributary joins a larger, still frozen river. Ice jams can be huge. The annual jam on the Liard River upstream of its junction with the Mackenzie in Canada's Northwest Territories, where the channel is about a kilometer wide, sometimes builds up to be 20 kilometers long and 4 to 8 meters thick; floating floes hurtle into its upstream end at speeds of almost 20 kilometers per hour.[14]

Sometimes an ice jam floats, allowing the river to continue flowing underneath it. Sometimes it is grounded, and so tightly packed that it functions as a dam; this causes flooding upstream. The floodwater overflowing the river's banks carries floes with it; they are left grounded on the floodplain if the floods recede before the floes melt (figure 8.1). Flooding can also happen downstream of an ice dam, as a result of the surge of water released if the dam gives way suddenly.

Floods bearing ice floes are even more devastating to floodplain dwellers than warm-weather floods. This is why the behavior of northern rivers at breakup is so closely studied. Investigations are difficult, because recording instruments placed to measure current speeds and temperatures are easily smashed by hurtling ice blocks. Nor can a river's discharge be easily predicted, because the relationship—assuming it is known—between stage and discharge at times of open water (see figure 5.3) does not hold at breakup time: whatever the discharge, the water level upstream of an ice jam is higher than it would be if the jam were not there.

Many northern rivers have a north-south alignment, and their behavior at breakup is strongly affected by their direction of flow. Breakup is comparatively gentle in south-flowing rivers: the ice at the southern (downstream) end melts first, and continued melting eats away the ice edge gradually, making the edge retreat steadily northward; bits of loosened ice float downstream into warmer, open water where they melt quickly without accumulating.

Things are different in a north-flowing river: the upstream reaches, where spring comes earliest, break up first, and the river, growing in volume and power, and laden with ice floes, bears down on the ice-bound reaches downstream, which are still in the grip of winter. Huge ice jams are apt to build up; floes carried in the fast current rear up, and crash into them. Such a jam is finally wiped out in a spectacular mechanical breakup.

Northerners living near a big river await the annual breakup at their particular spot on the river with keen anticipation, especially those who've made bets on the date. To prevent disagreements, gamblers have to decide precisely how they will define the instant of breakup, since the whole process, from the first visible rupture in a river's ice cover to the final disappearance of all the ice, may take many days. Therefore, breakup means whatever the local people say it means: at Lang Trading Post in the Mackenzie Delta, it means the time at which driftwood from the Peel River (a tributary) first floats past the trading post.[15]

8.4 Ice in the Soil

Invisible ice below the ground is as much a part of winter as the surface ice of frozen lakes and rivers. In the far north it is part of summer too: this is the vast region where much of the groundwater is frozen at all seasons; the frozen ground itself is known as *permafrost,* short for permanently frozen ground,[16] and it makes up about 60 percent of Canada and all of Alaska. As we shall see in the following section, permafrost is limited to high latitudes and it doesn't come right to the surface. But the surface soil freezes temporarily wherever there are frosty spells in winter; it happens even in Florida. Thus ice at and just below the ground surface is available, at least occasionally, for nearly all North Americans to inspect. It is the subject of this section.

Soil water freezes from the surface downward. When the air cools to below the freezing point on a winter evening, and the sun sets, the ground surface loses heat by radiation to the sky. Then the soil water at the surface crystallizes into ice crystals that grow downward. If moisture is plentiful and the cooling gradual, the result is *needle ice,* tufts of long, bristlelike ice crystals. The needles lengthen by expansion (water expands by 9 percent on freezing) as well as by growth (the crystallization of more and more water, added to the lower end of each needle), and the lengthening needles exert pressure, upward and downward; many tufts are forced up through the soil surface, carrying caps of soil on top. By morning the ground is crunchy with tufts of protruding needle ice that look like little brushes with dirt-tipped bristles. As the day warms, the tufts soften and bend over. Needles 5 centimeters or more in length are common; they are said to reach 30 centimeters occasionally.[17] Besides "sprouting" from the soil, needle ice can also grow on fallen logs and branches. Crystallization starts in the narrow crevice between dead wood and the loose bark encasing it, and the needles force the bark outward, opening up cracks in it as they grow.

Needle ice grows best where the atmosphere—this includes the soil atmosphere in the myriad soil pores—is very moist, and where the moisture freezes slowly at night and thaws frequently, if not every day then every few days. These conditions are common along riverbanks in regions too mild for rivers to freeze over in winter: moisture is ample, and the river itself acts as a heat reservoir, preventing nighttime temperatures from falling too far or too fast. The growth of needle ice, forcing up innumerable "crumbs" of sediment, causes considerable riverbank erosion; the crumbs roll down the slope into the water or are washed away when the river level rises. In areas with cold but not severely cold winters, this is the chief cause of riverbank erosion.[18]

Needle ice is one particular kind—an easily observable kind—of *segregation ice*. Segregation ice forms whenever water is sucked from moist soil and freezes onto earlier-formed ice particles. The growing ice doesn't always form needles. Often it develops into horizontal flakes known as *ice lenses* (more on these in section 8.6). But however the ice grows, it has marked side effects, both because it expands and because it brings about a rearrangement of the soil water. The expansion of ice in the soil is what causes frost heaving. The rearrangement of soil water happens because the ice must continually gain water from somewhere in order to grow. The needed water comes up from the unfrozen soil below the frozen layer which, besides providing the water, acts as a wick for moving it upward. In time the lower soil layer becomes dessicated. This is another reason why frozen soil crunches underfoot: the pore spaces in the dessicated layer are no longer held open by water and collapse under pressure from above.

The frozen layer at the surface sometimes remains thin even when icy weather lasts for many days. That is, the downward advance of the *freezing front*—the lower side of the frozen layer—slows to a standstill. This is good news for vegetable gardeners: it means that potatoes, beets, and winter turnips, for example, remain unfrozen for a surprisingly long time. The reason is that, as ice-cold water turns into ice, it gives off heat as it does so[19]—in other words, the very act of freezing releases energy. Thus there is often a temporary deadlock: the cooling caused by cooling weather and the warming caused by ice formation balance each other for a while, and the freezing front stays put. The phenomenon is called the *zero curtain effect*.

8.5 Permafrost

Permafrost means perennially frozen ground. The definition sounds straightforward enough, but it contains two ambiguities that have fueled much debate among experts.

First, how long a time should *perennial* imply? Ordinarily, permafrost remains frozen for centuries, but the borders of a tract of permafrost cannot fail to shift over small distances this way and that if the climate varies from time to time, as it always does, at least slightly. To avoid any indefiniteness, scientists have decided to apply the name *permafrost* to ground (soil or rock) in which the temperature has remained below 0°C uninterruptedly for 2 years or more.[20]

The second debatable point is, does permafrost always contain ice? The answer is, not necessarily. Solid bedrock totally devoid of water counts as permafrost if it obeys the rule about temperature; so does dry sand. But because it is unnatural to speak of a thing as frozen when it contains no ice, a technical term has been invented to cover the situation; the word is *cryotic.*[21] It means at a temperature of 0°C or less regardless of whether or not ice is present, whereas the word *frozen* (as applied to sediments or soil) means *containing ice.* The use of these words makes it easy to differentiate between perennially cryotic but ice-free ground on the one hand, and perennially cryotic, ice-containing ground on the other; the former is dry permafrost.

The term *wet permafrost* also makes sense. It is applicable to sediment that is perennially cryotic (colder than 0°C) but that contains liquid water. This is not a paradox: 0°C is the freezing point of *pure* water, and groundwater is seldom pure: it nearly always contains some dissolved minerals, and its freezing point is lowered by the presence of the minerals.[22] Indeed, that is the reason why, on the old-fashioned Fahrenheit scale, the freezing point of pure water is given by the seemingly arbitrary figure 32°; the temperature labeled 0° on that scale is the freezing point of a salt solution so constituted as to have the lowest attainable freezing point in the year (1709) the scale was invented.

So much for the untypical kinds of permafrost, which must be allowed for when we try to define the term precisely. Typical permafrost does contain ice. Sometimes it contains an excessive amount of it; that is, the pores in the ground contain so much ice that, if it were all to melt, the meltwater would be more than enough to fill the pores and some would be forced up through the ground surface.[23]

Having pinned down the exact meaning of permafrost, the next question to consider is, where is it found? Figure 8.2a shows the regions where it occurs in the Northern Hemisphere; small tracts (too small to be shown on the map) also occur at high elevations in the mountains, wherever the climate is cold enough. Notice how, on the map, the stippling thins toward its southern (outer) limit and finally peters out. This shows the way permafrost varies: as the section in figure 8.2b shows, permafrost underlies the ground surface almost everywhere in the High Arctic; the only exceptions are the ground below

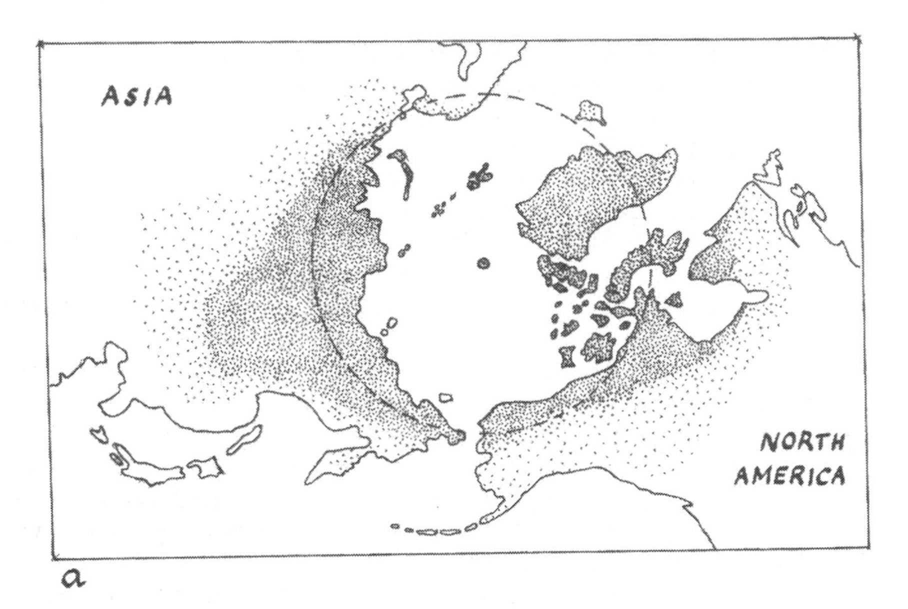

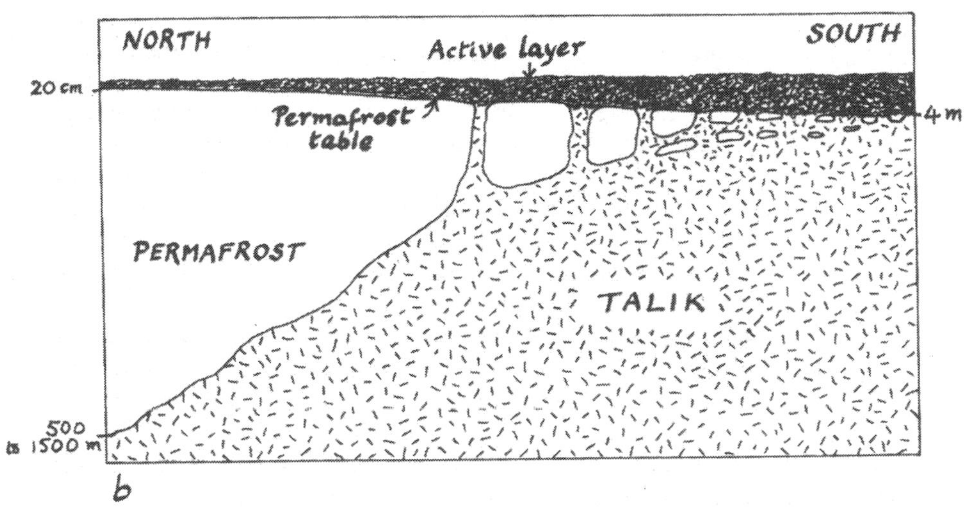

FIGURE 8.2. *(a)* The geographic range of permafrost in the Northern Hemisphere. The Arctic Circle is shown. Permafrost is uninterrupted where the stippling is dense; gaps become larger and more frequent toward the southern limit of permafrost, where the stippling fades out. *(b)* Cross-section showing how the thickness and continuity of the permafrost decreases, and the depth to the permafrost table increases, going from north to south (not to scale).

rivers and lakes too deep to freeze to the bottom in winter; under such lakes the heat stored in the water penetrates to a considerable depth and prevents the groundwater from freezing.

As you travel south, away from the rigors of High Arctic conditions, more and more gaps appear in the permafrost, beginning at the warmest sites, such as steep, south-facing hillsides where the local climate is unusually sunny and dry. Permafrost becomes more and more patchy the farther south you go through the subarctic; it is soon confined to cool, north-facing slopes, or to windswept uplands where not enough winter snow accumulates to insulate the ground and prevent whatever warmth it holds from radiating away. Continuing southward, the proportion of the land underlain by permafrost becomes less and less: the permafrost patches become smaller and more scattered, until finally they disappear altogether.

The depth and the thickness of permafrost vary too. The depth is the distance from the open air down to the top of the permafrost, known as the *permafrost table;* it ranges from about 20 centimeters in the High Arctic to 4 meters or more near the southern limits of permafrost.

The thickness of the permafrost itself depends on the coldness of the climate, and on how long the cold has lasted. It takes thousands of years for groundwater at great depth to respond to temperature changes at the surface, and the deepest layers of present-day permafrost must have frozen several thousand years ago, some of it when the huge ice sheets of the last ice age were in existence. Modern permafrost is much thinner in areas that were then covered by ice sheets (most of arctic Canada) than in areas that were not covered (parts of Alaska and arctic Siberia); the explanation may be that an ice sheet functioned as a blanket during ice-age times, insulating the ground below it from the worst of the cold.[24] The thickest permafrost now known is found in Siberia, and is about 1,500 meters thick.[25]

Now for the ground in permafrost country that *isn't* continuously frozen: next to the surface is the *active layer,* ground that thaws in summer and freezes in winter. Elsewhere are blocks and zones of *talik,* ground that remains unfrozen (strictly speaking, *noncryotic*) all through the year.

Talik is found in various locations (figure 8.3). Throughout permafrost country, the ground below the bottom of the permafrost is commonly spoken of as talik, although it is no different from ground at the same depth in any latitude, which isn't called talik. Its warmth comes from the heat of the earth itself, because of which the temperature of the ground tends to rise at increasing depths. The rate at which it rises varies from place to place but is usually

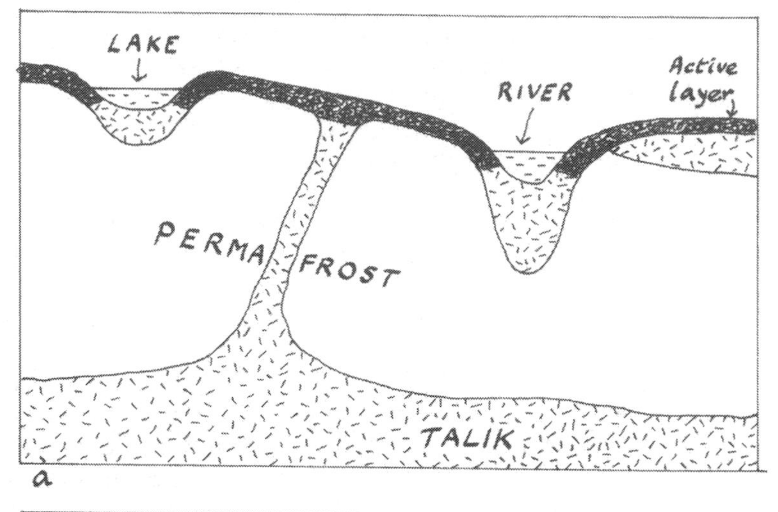

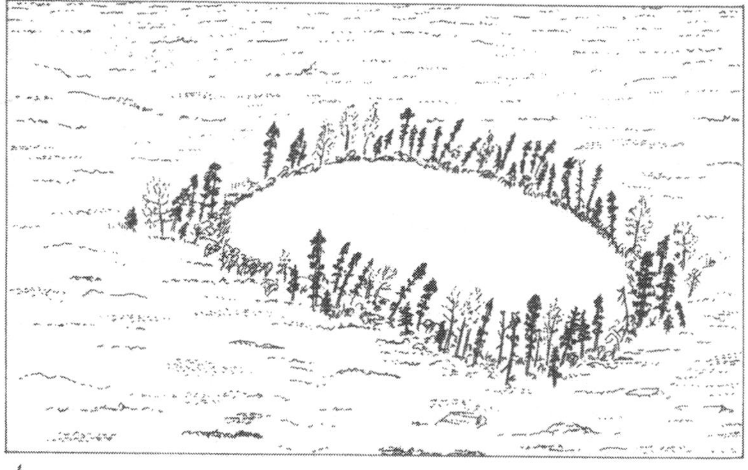

FIGURE 8.3. *(a)* Cross-section showing different kinds of talik: above the permafrost are taliks under a lake and a river, and a suprapermafrost talik. A hydrothermal talik links the deep talik below the permafrost with the active layer. Note that the scale of the section is about 1,000 times greater than that of figure 8.2b; the latter is on too small a scale to show lake and river taliks. *(b)* Aerial view of a deep tundra lake, surrounded by trees (black spruce and larch) 10 meters tall at most. Trees cannot grow elsewhere, as the soil is too shallow to anchor them.

between 2°C and 3°C per 100 meters. This "layering" of the ground temperature is what controls the thickness of the permafrost: its lower surface is at the level where the temperature first rises (as you go downward) from below to above freezing point. The groundwater in this deep talik is believed to be connate water,[26] that is, ancient water that has been there since the rocks were first emplaced (see section 2.1).

The "warm" ground below those rivers and lakes too deep to freeze to the bottom in winter is also talik. Such taliks (the word is used in the plural for separate blocks of talik, as well as in the general sense) may be isolated pockets, or else they may penetrate right through the permafrost, opening into the widespread talik below.

Other taliks include suprapermafrost talik, which sometimes forms a zone at the bottom of the active layer in latitudes where the permafrost table is deep.[27] There are also hydrothermal taliks, which are kept from freezing by slowly moving groundwater tunneling a route for itself up through the permafrost from below; the rising water is, in a sense, a "hot spring," even though its temperature may be barely above freezing. If it emerges at the surface as a seep, the water seeping in winter instantly freezes into a layer of aufeis,[28] in the same way that water leaking through cracked river ice does. Layer upon layer of aufeis sometimes builds up over a seep, into a high mound of ice.

Groundwater in "permafrost country" threads its way through these taliks in the same way that groundwater in a warm climate moves through aquifers between confining layers. Indeed, waterlogged talik *is* an aquifer; so, in summer, is the lower part of the active layer, as we shall see below.

Permafrost forms the confining layers, and, like other confining layers, it is seldom wholly impermeable; that is, its hydraulic conductivity, though very low, is seldom zero. This is because permafrost usually contains at least a small amount of liquid water. As groundwater freezes, the minerals dissolved in it are excluded from the ice as it forms, and they become more and more concentrated in the water yet unfrozen; this unfrozen fraction—a solution of mineral chemicals—eventually becomes too concentrated to freeze at the ambient temperature. That accounts for some of the liquid water. The rest is hygroscopic water (see section 4.6), the water that clings as an exceedingly thin film to the surface of every tiny particle of rock in wet sediment; this water can supercool to as low as −50°C without freezing. The amount of it in fine-grained sediments is often considerable; for example, 20 percent of the water in frozen clay is sometimes supercooled hygroscopic water.[29]

The presence of liquid water makes permafrost capable of conducting water to some degree: this means that it is most often a slightly leaky confining layer, though the "leak" is normally exceedingly slow.

Now consider the active layer that forms the surface of the land over vast tracts of the arctic tundra—the so-called barrenlands. In summer much of the active layer is waterlogged. The waterlogged parts are aquifers, with the almost impermeable permafrost acting as the confining layer below them. It's a remarkable situation when you think of it. Going downward from a point not far below the surface (sometimes from the surface itself), all the pores in soil and bedrock are completely filled with water. Layers where the water is liquid are aquifers; layers where it is solid (as ice in the permafrost) are confining layers: it is the state of the water—whether liquid or ice—that governs how each layer shall behave. Climatic warming or cooling can convert a confining layer into an aquifer or vice versa.

When summer warmth melts a gently sloping active layer, the water naturally drains downhill. As it drains, it erodes the surface it is flowing across—the permafrost table—making subsurface valleys for itself; in other words, subterranean erosion creates subterranean topography. This odd phenomenon happens because flowing water can erode frozen ground extremely rapidly, by melting the frozen soil and then washing it away (this is thermoerosion; see section 6.4). The flow of meltwater along the subsurface valleys so formed may be greater than the flow over the surface.[30] The valleys are known as *water tracks,* and their invisible presence is sometimes shown by visible bands of wetland vegetation on the tundra, and sometimes by "rivers" of rocks, left high and dry because the soil in which they were imbedded has been rinsed away.

In level tundra with a very wet active layer, the water table in summer is above ground in many places, causing the landscape to be dotted with innumerable shallow lakes. These are *thaw lakes,* and they occupy every slight depression in the ground. They range in size from a few hundred meters to several kilometers long, and often occur in closely packed swarms, all aligned in the same direction,[31] probably by the wind. They are shallow—1 or 2 meters deep—and freeze to the bottom in winter. One of the ways in which these depressions are often formed is described in the next section.

8.6 Subterranean Ice

Most of what could be called frozen groundwater is dispersed through the permafrost as particles and small lenses of ice in the pores, cracks, and crannies of the ground: but not all. Some of it exists as masses of pure ice.

Probably the largest masses are buried sheets of segregation ice. As we saw in section 8.4, segregation is the process by which ice lenses form: water is sucked upward by capillary action, the force that makes water rise in a sponge; it migrates to the freezing front and there freezes. If water is ample and temperatures remain below freezing for long periods, the ice lenses unite as a flat sheet, which keeps on growing thicker and thicker as more and more water freezes onto the bottom: sheets as much as 40 meters thick have been found. Their tops often form the floor of the active layer. The ice may be clean and white or grubby with included fine sediment.

Buried ice sheets can last a long time. Some sheets near the Mackenzie Delta are known to be at least 40,000 years old. Their age is inferred from the fact that they are sandwiched between datable layers of sediment. Moreover, the ice layers and the sediment layers have been faulted and folded as a unit, with the component layers held parallel; this shows that the ice could not have become segregated after the faulting and folding happened: it must have been frozen, uninterruptedly, since before that time.[32]

Not all buried layers of ice are segregation ice. Some are surviving remnants of the ice sheets of the last ice age, which had all vanished from the North American mainland by 6,000 years ago at the latest. Remnants were able to survive if they happened to acquire an insulating covering layer—a thick blanket of windblown dust, perhaps, or a mudflow from nearby hillsides, or a sediment layer deposited by streams of meltwater from the ice front. Still other buried ice layers are frozen lakes that happened to become sediment-covered.

Not all buried ice consists of horizontal sheets. Along the coastal plain of the western North American Arctic are innumerable *pingos,* small (rarely more than 50 meters high), steep, conical hills scattered over the level plains like little volcanos (figure 8.4). Under an outer veneer of soil, clothed with ordinary tundra vegetation, they are composed of solid ice.[33]

A pingo grows when a tundra lake drains, leaving exposed a pocket of waterlogged talik that no longer has the lake above it as insulation. The talik becomes trapped and compressed between a "floor" of old permafrost and a "ceiling" of new permafrost created by the newly frozen lake bed: the old permafrost table keeps rising, owing to the disappearance of the lake, while the new freezing front descends within the erstwhile lake bed, which has begun to freeze from the top downward. The talik is thus caught between rigid surfaces growing downward from above and upward from below; the water in it is squeezed out at the top, and freezes—and in so doing expands, as water always does (by 9 percent) on freezing. The result is a lump of ice that forces up the frozen layer over it.

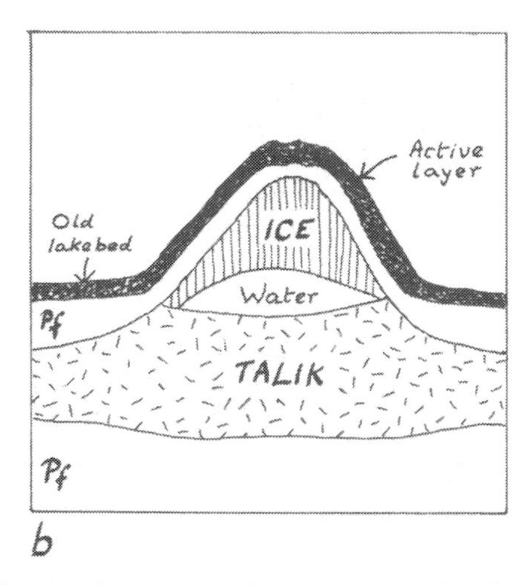

FIGURE 8.4. *(a)* Aerial view of a large pingo, with an eroding summit. The numerous ponds and the river in the surrounding plain are remnants of the lake whose draining caused the pingo to grow. *(b)* Cross-section through the pingo; see text for details. *Pf,* permafrost.

The "floor" and "ceiling" continue to grow toward each other, the moisture in the sediment between them continues to be squeezed out, and then to freeze and expand, until all available water is used up. At this stage the pingo has reached its maximum size and its core is a conical plug of pure ice. Later, erosion, accompanied by melting, will wear the pingo away and it will disappear. Pingos are not widespread in the Arctic; they grow only where the sediment texture and the climate are exactly right, and when small lakes deep enough to have taliks beneath them happen to become drained. A large pingo is believed to take about 1,000 years to grow to its full height.[34]

The most widespread examples of buried ice in the Arctic are *ice wedges* (figure 8.5). Over large tracts of tundra, rapid cooling in winter, to very low temperatures, causes the soil to contract and crack: the cracks form extensive networks. Every summer, when the active layer thaws, meltwater drains into these long, narrow, vertical cracks, and every fall, when subzero temperatures return, the water in the cracks freezes into vertical seams of ice. The seams create lines of weakness in the soil, causing the cracks to open in the same place year after year. Thus the ice in each crack builds up in vertical annual layers, becoming an ice wedge. A wedge may keep on growing for centuries or millennia; big ones are as much as 10 meters across at the top and more than 10,000 years old.

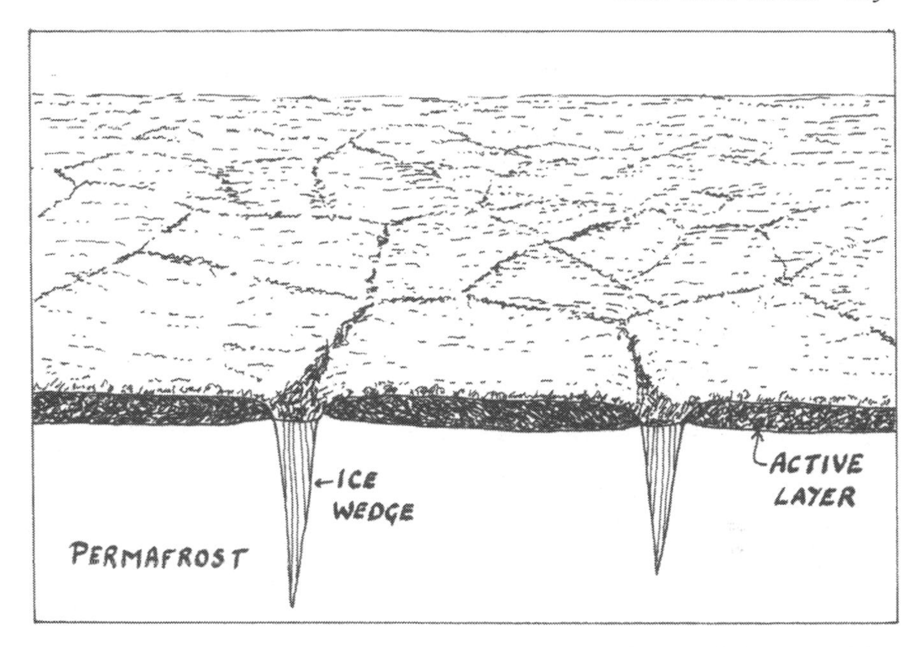

FIGURE 8.5. A field of tundra polygons. The cross-section in the foreground shows two of the ice wedges underlying the troughs between the polygons

An ice wedge is normally buried and invisible, though its presence is usually revealed by a straight, shallow trench at the surface; the ice itself can be seen if the wedge is cut by a riverbank, or a coastal bluff where soil slumping, which often happens at such sites, has left it exposed in cross section.

Networks of ice wedges often partition a big expanse of tundra into a pattern of *tundra polygons* (figure 8.5). The polygons—they are often rectangular—may be 50 meters or more across, too big to be easily recognized as such by anybody standing on the ground but a conspicuous part of the scenery from the window of a plane. The trench along the top of an ice wedge often has raised banks on either side, where the ever-widening wedge has forced the soil up to left and right of it. A polygon surrounded by such banked trenches is therefore a shallow basin, where the high water table of summer is likely to lie above the surface, creating a shallow pond. An array of such ponds is often the preliminary stage in the development of a thaw lake. Each pond, however small, grows in area by eroding its shores; thermoerosion is so efficient that even the tiniest ripples have an appreciable effect, and as a pond's area increases, tundra winds raise bigger waves, capable of stronger erosion. Before long, adjacent ponds coalesce with one another, and this is one of the ways thaw lakes come into existence.[35]

Underground ice, though seldom directly visible, thus has a marked effect on the scenery: it forms all but the outer skin of pingos, and it is the indirect cause of many of the innumerable lakes that dot the summer tundra. It mustn't be overlooked, however, that even though about 77 percent of all the world's fresh water exists as ice, only a very small fraction of that ice (how much is unknown) is underground. By far the greater part is above ground, as ice sheets and glaciers, and they are another story.[36]

9

Dams, Diversions, and Reservoirs

9.1 Adapting Natural Waters for Human Purposes

Earlier chapters considered wild lakes and rivers, as they are found in nature. In this chapter we consider the tamed variety. Humanity has tinkered with natural fresh water for thousands of years. Dams are among the biggest and most spectacular structures ever built by mankind. The earliest dam that history records was built in Egypt, about 4,500 years ago; it was an earth dam, as all dams were until cement-based concrete was invented, and was built across the channel of a river that flowed only in the rainy season. It didn't last: unusually heavy rains one year washed it away.

Modern concrete dams exist in the thousands. North America's big three are the Hoover Dam across the Colorado River near Las Vegas, completed in 1936, which has created Lake Mead; the Grand Coulee Dam across the Columbia River in Washington, completed in 1942, which has created Lake Franklin Roosevelt; and Shasta Dam across the Sacramento River, completed in 1945, which has created Lake Shasta. Their heights are, respectively, 221 meters, 168 meters, and 183 meters; their lengths across the top are, respectively, 379 meters, 1,272 meters, and 1,055 meters. The corresponding measurements for the short-lived Egyptian dam of 4,500 years ago are 11 meters and 114 meters. It is too early, of course, to judge what the life spans of Hoover, Grand Coulee, and Shasta dams will be, but nothing is eternal, and nobody knows whether they will outlive the enthusiasm for engineering megaprojects, among them big dams, that seems to have peaked (at least in the so-called developed nations) sometime in the middle of the twentieth century. Hoover Dam has been described as "the greatest structure on earth, perhaps the most significant structure that has ever been built in the United States."[1] According to the same source, Grand Coulee Dam, though much lower than Hoover Dam, is a larger structure than Hoover and Shasta dams put together. When they were built, these engineering marvels were regarded with awe.

The undesirable consequences of building dams and diverting rivers are becoming more and more apparent as the twentieth century draws to a close. Before discussing the drawbacks, however, one must bear in mind the virtues of dams—the reasons they were constructed in the first place.

The main reasons for damming a river are to provide hydroelectric power; to facilitate navigation; to create a reservoir, for a city's water supply or for the irrigation of agricultural crops; and to control flooding. The two last-named purposes are sometimes at odds: a reservoir needs to fill up in the rainy season if it's to supply water in the succeeding dry season, but it needs to be empty in the rainy season if it's to serve as a protection against floods. A full reservoir sometimes becomes the source of catastrophic floods if its retaining dam collapses when the reservoir is full.

Irrigation of arid land is the primary purpose of many dams, in spite of the fact that evaporation from the surface of a large reservoir leads to a net loss of water from the watershed in which the reservoir lies. By allowing the surviving water to be channeled to where it is wanted, irrigation has allowed vast areas in the drier parts of the world to be made agriculturally productive, and in this way has increased the world's carrying capacity for humans to an artificially high level, putting off, temporarily, the ultimate collapse that uncontrolled population growth must inevitably bring. But this increased carrying capacity—the total number of humans the planet can support—will not last. Irrigation damages land: it causes excess salts to accumulate in the irrigated soil, which eventually becomes unproductive and unusable.[2]

The advantages of irrigation therefore have a price. So have all the other benefits to be gained from interfering with rivers—damming them, diverting them, and channelizing them. Unnatural waterways and their characteristics are the topics we now consider.

9.2 *Reservoirs*

For people living in regions not blessed with natural lakes, reservoirs provide substitutes. Vacationers who treat a reservoir as an artificial lake and go there for swimming, boating, and fishing are often unaware of the unnaturalness of their playground until the dam's managers lower the water level, leaving resorts, cottages, and camps separated from the water by a wide tract of slowly drying mud. In forested country a reservoir often has a belt of drowned trees emerging from the shallow, near-shore water, making it an unattractive substitute for a real lake.

Many of the differences between reservoirs and true lakes are less obvious. For example, the average reservoir has a much longer shoreline than the average lake of the same area;[3] this is because the basin of a reservoir, unlike that of a lake, is typically a long, narrow river valley and includes, branching out on either side, the lower reaches of its tributaries' long, narrow valleys. The deepest point in a reservoir is typically at one end, just behind the dam that impounds it; the deepest point in a lake tends to be somewhere near its center.

The residence time of a reservoir (the time a given drop of water remains in one) is usually shorter than the residence time of a lake. When the residence times of several reservoirs in Kansas were compared with those of an equal number of natural lakes of similar size in Michigan, it turned out that the average residence time was about 14 months in the reservoirs as against nearly 4.5 years in the lakes.[4]

The most important differences between reservoirs and lakes, however, are in their chemistry, and arise from the fact that creation of a reservoir entails the flooding of land and the submergence of vegetation. Newly filled reservoirs usually contain large amounts of dead plant material, which serves as a habitat for rapidly multiplying bacteria capable of absorbing any mercury that happens to be in the soil; they convert the mercury into a form (methylmercury) that fish can ingest, and in this way mercury enters the food chain. Methylmercury is about 100 times more toxic than the original mercury compounds in the soil, and it becomes steadily more concentrated—biomagnified—as it moves up the food chain to the top predators, fish: by the time it reaches the fish, the concentration may have increased a millionfold.[5] Mercury pollution is much commoner in reservoirs than in natural lakes because the drowning of land vegetation creates the habitat required by the bacteria. It makes no difference whether the reservoir is a wholly new "lake" or an enlarged version of a preexisting lake that has been dammed at its outlet.

Increased mercury in fish means increased mercury in people eating the fish, with serious consequences for indigenous populations that depend on locally caught fish for a large fraction of their diet. The Crees of the James Bay region of Quebec are an example.[6] The natural lakes from which they obtained their fish were converted to reservoirs in the course of developing the region's hydroelectric power, whereupon the mercury level in the fish rose steeply: it increased fivefold in samples of northern pike (the topmost predator below humans), which brought it to six times the legal maximum concentration for commercial fish in Canada. At the same time, the mercury level in hair samples from many of the local Cree rose above the 6 parts per million (by weight) recommended as an upper limit by the World Health Organization. The quantity varied greatly from person to person, and there is much disagreement over what should be regarded as a level dangerous to health, because distinguishing the effects of chronic, low-level mercury poisoning from other afflictions is difficult. High concentrations cause Minamata disease, marked by irreversible damage to the brain, liver, and kidneys, and finally, death.

Reservoir construction has also caused mercury pollution in northern Manitoba, where several lakes have been enlarged by damming their outlet.[7] The amount of mercury in fish muscle rose immediately the land was flooded and continued to rise for 2 or 3 years, after which it showed little sign of declining. The result was the collapse of a valuable commercial whitefish fishery.

Indeed, mercury poisoning has dire consequences for all the animals of an affected ecosystem (and possibly the plants too); it can cause reproductive failure, blindness, and brain damage that modifies an animal's behavior in ways that threaten its survival. Fish-eating humans are merely the top predators in such an ecosystem.

Another trouble with reservoirs that contain masses of decomposing vegetation is that they release to the atmosphere large amounts of the two chief greenhouse gases, carbon dioxide and methane. If the decomposing material is peat—compressed plant remains that have been accumulating for centuries—the release of greenhouse gases is likely to continue indefinitely or, at any rate, for as long as the reservoir lasts. A reservoir powering a hydroelectric plant can sometimes give off as great a quantity of greenhouse gases as a coal-fired generator yielding the same electric power,[8] so it is a fallacy to think of hydro power as "clean."

Reservoirs are more than "defective lakes": they are massive bodies of water collected in "basins" not designed by nature to hold them. The tremendous weight of water in a large reservoir deforms the earth's crust beneath it, sometimes breaking it: the result is an earthquake, which can exceed a magnitude of 6 on the Richter scale.[9]

The amount of water now stored in reservoirs worldwide is enormous. By the end of the twentieth century it will probably exceed 10,000 cubic kilometers;[10] spread on land (and forgetting that water flows downhill) this would give a layer of water 10 centimeters deep, covering all the dry land in the world. Withdrawal of all this water from circulation is estimated to have lowered the world's sea level by about 3 centimeters. It is difficult to infer the net effect on sea level, as it is hard to judge to what degree the drop caused by storage in reservoirs compensates for the rise caused by global warming, which is increasing ocean volume. The increase has two causes: expansion of the water as its temperature rises, and added water coming from melting ice sheets and glaciers.

The enormous weight of the water held back by dams is affecting the earth's rotation.[11] A great weight has been shifted from the equatorial part of the ocean, where it was a long distance from the earth's axis of rotation, to reservoirs in midlatitudes, where it is much closer to the axis. As a result the earth's rotation has speeded up, in exactly the same way as a skater spinning with outstretched arms spins faster if she draws in her arms close to her body. The effect is negligible in comparison with that caused by various other factors, the drag of ocean currents for example, but it is nevertheless real; in less than half a century, the "reservoir effect" has shortened the length of the day by nearly 0.00001 second.

Moreover, because the reservoirs are not symmetrically located around the earth's axis, the axis itself has tilted. In other words, the endpoints of the axes, the North and South Poles, have shifted; each is now 60 centimeters away from the point it occupied before the reservoirs were filled. Of course, the equator has shifted too, maintaining its position halfway between the poles; the latitude and longitude of every point on earth has changed correspondingly. To be fair, these shifts are not as great as those caused by natural "polar drift" that goes on all the time. But they do emphasize how delicately the earth is poised, and how illusory the fixedness of a fixed point really is.

9.3 The Many Consequences of Damming a River

The presence of a dam profoundly alters a river valley, both above the dam and below it. We'll consider these areas in turn.

Above a dam, two ecosystems are destroyed, sometimes more: the two that are automatically destroyed are the land ecosystem submerged by the reservoir, and the original river ecosystem now engulfed in the reservoir's standing water.

The submergence of a land ecosystem entails the drowning of all air-breathing organisms and, frequently, the destruction of homes, gardens, farms, and whole villages. Often the ancestral lands of indigenous people simply disappear—"lands" become "waters," against the wishes of the people who have lived there for generations.[12]

The destruction of the aquatic ecosystem that previously occupied the river upstream of a dam often goes unremarked because it is invisible; but it happens. Flowing water, well-oxygenated and well-lighted, is replaced by the still water at the bottom of a reservoir, poorly oxygenated and dark. Few living things can endure so radical a habitat change; merely being an "aquatic organism" is not enough to assure survival.

Upstream of a reservoir, the original river and its tributaries still flow, but in an altered state. Their flow is slowed, so that sediments that would have been carried downstream are deposited, burying the riverbeds and leaving their channels shallower and narrower. This in turn makes flooding likely in places where floods were unknown before.

A case is known of a big dam that, surprisingly, protects a natural ecosystem (the effect was unintentional!). Mica Dam was built on the Columbia River in British Columbia in 1973; the reservoir behind it, called Kinbasket Lake, fills a long stretch of the Columbia valley and has submerged the highway that used to parallel the river. Disappearance of the highway has made the forests on the valley slopes difficult to reach and, consequently, uneconomic to log. The forests will probably not remain protected for long, however; a new road is to be built.

Now for the downstream effects of a dam: the best-known effect is that on fish. First, consider *anadromous fish,* the ones that live most of their lives at sea but swim up rivers to spawn; the economically important ones are salmon, both the Atlantic and the Pacific kinds. Anadromous fish swimming up river are stopped short of their spawning grounds by a dam unless something is done to help them, and that something—either fish ladders or a ferry service around the obstruction—is often the only action taken to mitigate the undesirable effects of a dam: consequences that don't affect commercial or sport fisheries, or don't affect them in an obvious way, tend to be disregarded, unless the public is sufficiently informed to make a fuss.

The state of the river downstream of a dam depends on how much water is allowed to flow through or around it. If the dam's only purpose is to supply hydroelectricity, the discharge downstream may be no less than it was before, albeit altered. The surplus water that has to be disposed of when the demand on the generators is low often flows too fast for the well-being of fish that happen to be in the line of fire. The water plunging over a spillway carries entrained bubbles deep into the plunge pool below, sometimes so deep that the pressure is enough to make the bubbles dissolve: the water becomes supersaturated with dissolved air, and fish swimming in it die of gas-bubble disease—the bends.[13]

Irrigation dams exist to capture water and divert it; excess water allowed to flow through a dam into the downstream river channel is usually released from the reservoir's lowest point. If the reservoir is narrow and deep, the water is likely to be strongly stratified in summer, with the deepest level deficient in oxygen and very cold; the river ecosystem that receives it is then at risk of being damaged or destroyed.[14] When a sudden water release causes a downstream fish kill,[15] as often happens, it is reasonable to conclude that the water was ecologically "faulty," and that other components of the ecosystem have been affected as well as the fish.

In other circumstances the much-reduced river below a dam may be so shallow, and so sluggish, that the water becomes too warm for salmon and trout. It may become weed-choked. The flow may lack the force to scour debris and sediment away from spawning gravels in advance of spawning,[16] and the gravels become clogged and useless. The river may even, at times, be reduced to a string of isolated, stagnant pools. Tributaries entering such a diminished river drop their sediment loads on entry, making the river still shallower.

Altering a river's regime—the way its flow varies seasonally—inevitably affects all the organisms, plant and animal, adapted to its natural fluctuations. In section 6.10 we considered the effect of dams on pulse-stabilized river-valley ecosystems whose very existence depends on spring flooding. Disturbance of a river's regime also affects fish: for example, the white sturgeon living in the big rivers of the Pacific Northwest are believed to spawn in the strong currents of spring runoff.[17] Hydroelectric dams, particularly those on the Nechako, Kootenay, and Columbia Rivers, block the normal spring freshets and hold back the water until it is most needed for generating electricity, that is, the following winter. The sturgeon fail to spawn, and young ones are now scarce in British Columbia; they are outnumbered by older fish, born before the dams were built.

The cessation of a river's normal spring flooding sometimes does damage that extends to coastal marine waters. The brackish water ecosystems of deltas and estuaries, and the marine ecosystems just offshore, are nourished by nutrients brought in by rivers; when a river is dammed, nutrient-rich silt is trapped behind the dam instead of being carried out to sea where it is needed. Nutrients also originate in the reaches of the river below the dam, but the river's current may be slowed so much that nutrient-bearing sediments are deposited before they reach the sea. Indeed, "regulation" of a river, by evening out the inequalities of the river's flow over the seasons, means that the nutrients that normally come with a rush, in time for the annual burst of biological activity, fail to arrive; instead, they are left to accumulate unused at the bottom of a river whose flow is kept unnaturally gentle at all times. This phenomenon has caused the collapse of many coastal fisheries.[18]

Finally, consider the consequence of damming a north-flowing river in a cold, northern climate: the unnatural slowing of the river's flow may delay spring breakup and thus delay spring itself for all the life in the water.

9.4 Channelization and Diversion

Rivers are mistreated by channelizing them and diverting them, as well as by damming them. Channelization and diversion both have the effect of increasing a river's rate of discharge (equivalently, speeding its flow), which may be as undesirable as decreasing the discharge (slowing the flow).

Channelization consists in artificially "tidying" a river's channel by straightening and smoothing it;[19] the object is to allow unusually heavy flows to drain away fast, and so to prevent unwanted floods. Not surprisingly, a river's natural ecosystem is drastically altered by channelization: pools and riffles are eliminated; so are meanders, with their alternation of shady, undercut banks and shallow reaches. Smoothing of the channel removes living plants and submerged logs, snags, and boulders. In brief, a river's naturally high diversity of habitats is greatly reduced. The faster flow speeds up erosion, making the water muddy and increasing sedimentation where the flow returns to normal downstream of the channelized section. The frequent maintenance operations needed to control the extra erosion prevents the development of a new community of plants and animals—a new ecosystem—adapted to the changed conditions. Fish can no longer retreat to calm backwaters when the flow becomes too much for them, and their eggs are at risk of being swept away. These are the unwanted side effects of channelizing a river, and they should always be carefully considered and evaluated before engineering work proceeds on the channelization of a flood-prone river.

The consequences are similar, though much more dire, when a river is diverted into the channel of another river: two rivers are compelled to flow where one flowed before, and tremendous forces are let loose if the rivers are large. An example is the diversion of the Churchill River into the Nelson River in northern Manitoba, southwest of Hudson Bay. According to scientists familiar with the diversion, "the instabilities created in the environment are essentially beyond control."[20]

Combining the waters of two rivers seriously endangers the ecosystems—including stocks of commercially valuable fish—in both, because new predators, new diseases, and new parasites may enter water hitherto free of them.[21] In the normal course of events it is almost impossible for fully aquatic organisms, incapable of surviving out of water, to get from one watershed to another. Each watershed in a region is biologically isolated from the others, and its plant and animal populations become mutually adapted; they live as a self-contained community in lasting equilibrium. If the isolation is destroyed, so is the equilibrium; new predators and disease organisms invade and can wreak havoc in the affected ecosystems. What happens resembles the decimation of indigenous peoples, in many parts of the world, when European explorers arrived bringing new epidemic diseases with them.

10

Wetlands

10.1 Wetlands: What, Where, and Why

North America's surface fresh water (indeed, all surface fresh water) is contained in lakes, rivers, artificial reservoirs—and wetlands. Wetlands are sometimes treated as an afterthought: in fact, they are of tremendous importance in the scheme of things, from the hydrological, ecological, and biological points of view.

North American wetlands cover a truly vast area; for example, about 15 percent of the total area of Canada[1] and 3.5 percent of the coterminous United States[2] are wetlands. They are more productive, in terms of plant growth, than either agricultural land or natural grassland.[3] They are storehouses of biodiversity. They are valuable to humanity as water-storage sites, holding back floods in the wet season and gradually releasing the water later, in times of drought. They filter out pollutants and sediments from the water that passes through them. They are the irreplaceable habitat for vast numbers of ducks, wading birds, and shorebirds, who breed there or stop there to feed while migrating. We should be thankful for their existence, and do all we can to protect and preserve them.

Wetlands form wherever flat, poorly drained land collects enough water—from precipitation or runoff—for the surface to be submerged or saturated much or all of the time. This may seem vague: if a wetland remains a wetland even though it is sometimes dry, how is it to be recognized as a wetland? The answer is that the land must be wet enough, or flooded often enough, to support typical wetland vegetation, which is unmistakable and differs conspicuously from the vegetation of well-drained land. In a word, a wetland is defined by its vegetation, not its hydrology.

Wetlands form in a multitude of places; any depression in the ground that is too shallow to serve as a lake basin will be a wetland if it is poorly drained. Wetlands form on frequently inundated river floodplains and in the lee of artificial levees built to prevent rivers overflowing their banks. Lakes that become shallow as they gradually fill up with the remains of dead aquatic plants, and the sediments trapped by the plants, eventually develop into wetlands.

There doesn't have to be a depression in the ground for a wetland to form, however.[4] It can happen where the surface is kept constantly wet by a row of seeps along the base of a hill; the water seeps out where the sloping water table under the hill meets the soil surface. And, as we shall see, some wetlands are capable of expanding beyond the boundaries of the depressions they occupied initially.

Wetlands are particularly abundant in regions having an immature drainage system, that is, where the drainage system is incompletely developed. This is true of the land that was covered by thick ice sheets during the last ice age (see figure 3.2). The 10,000 years or so[5]—depending on the location—since the ice melted have not been enough for streams and rivers to erode a continuous, linked system of channels draining all the once-glaciated ground to the sea. Much undrained or ill-drained land still remains and is the site of numerous wetlands. This is why wetlands occupy so much more land in Canada than in the United States: nearly all of Canada was ice-covered.

Now to introduce the different kinds of wetlands. The variety is tremendous. As a broad classification, there are *bogs, fens, marshes,* and *swamps,* but each of these wetland types includes a large number of recognizably different subtypes. For example, among the numerous subtypes of bogs are domed bogs, blanket bogs, plateau bogs, flat bogs, and basin bogs; fens, marshes, and swamps are each classified into several different subtypes too.[6] Shallow water also rates as wetland rather than as lake or pond if the depth of the water in the dry season is less than 2 meters.[7] Shallow water differs from other kinds of wetland in having a relatively large amount (75 percent, to be precise) of open water clear of vegetation. A beaver pond is a shallow-water wetland while it is maintained, and usually continues as a wetland of some other kind for years after the beavers who made it have moved away.

The following sections describe what makes wetlands as a whole so varied. The differences are mainly due to the duration of the annual dried-out period (if there is one), and on whether their water is stagnant or gently flowing.

10.2 Peatlands: Bogs and Fens

Two of the principal kinds of wetlands are bogs and fens; together, they are known as *organic wetlands, peatlands, mires,* or *muskeg.* (Students of wetlands have come up with a remarkably rich vocabulary of special terms.) Peatlands, as the name implies, are underlain by a thick layer of peat. To count as a peatland, the layer must be 40 centimeters or more thick;[8] then one can be sure that almost all the plants growing there are rooted in the peat.

Peat is a very special material. Pure peat is entirely organic, containing no mineral matter, such as sand grains or clay; it consists of nothing but dead plant remains, incompletely decayed. It forms when the remains of dead plants accumulate, year after year, in cold water, where the low temperature and lack of oxygen inhibit the activities of the bacteria that normally decompose plant remains. Peatlands develop best in a cool climate, and they develop slowly. It takes a long time for quantities of dead plant material to accumulate where plants grow slowly because of a short, cool growing season; a big tract of peatland is often centuries old.

A peatland is permanently wet at depth, though its surface sometimes dries temporarily in hot summer weather. The way the water behaves determines whether a peatland shall be a fen or a bog. The water in a *fen* is groundwater seepage or slowly flowing surface water, and has therefore picked up some mineral nutrients to nourish the vegetation; thus fen water moves, albeit very slowly. The water in a *bog* is largely rainwater; it is stagnant, acidic, and exceedingly poor in nutrients.

These differences make a spectacular difference to the vegetation; fens are commonly covered with grassy plants, chiefly sedges, which form a uniform, level expanse of green at the height of the growing season. The plants need mineral nutrients and neutral or mildly alkaline water. Bogs are covered with plants (figure 10.1) specially adapted to grow in the unpromising conditions: the commonest is peat moss (various species of *Sphagnum*), accompanied by a variety of low shrubs such as bog rosemary, bog laurel, Labrador tea, cranberry, and leatherleaf. Two species of tree, black spruce and tamarack, also do well in bogs. Some bog plants are carnivorous, for example, pitcher plant

FIGURE 10.1. Some common bog plants (not to scale). *(a)* Labrador tea, *Ledum groenlandicum* (flowers white); *(b)* bog laurel, *Kalmia polifolia* (rose-purple); *(c)* sundew, *Drosera anglica* (white); *(d)* pitcher plant, *Sarracenia purpurea* (maroon); *(e)* peat moss, *Sphagnum* (whole plant green or purplish red). Sundew and pitcher plant are insect eaters.

and sundew; they compensate for the nutrient shortage by capturing and ingesting insects.

Bog plants tend to be more colorful and attractive than fen plants, and, unlike grasses and sedges, many have beautiful flowers. Bogs are less uniform in appearance than fens. They often have a bumpy surface because of the way sphagnum grows in hummocks, and their other vegetation is more varied. Their coloring is more variable too, as it depends on which species of plants occupy the most space: the several different sphagnum species range in color from pale green to reddish purple, and many of the shrubs have bronze leaves. White tufts of cotton grass often dot the scene.

The peat is not the same in bogs and fens: fen peat consists of the remains of sedges and mosses (excluding sphagnum), and bog peat of the remains of sphagnum and other bog plants. The difference is obvious when the two kinds of peat are dug up and examined carefully. Also, fen peat tends to be more decomposed than bog peat because it accumulates in slowly moving, and therefore better-oxygenated, water.

It's worth repeating, because it is often misunderstood, that the difference between a fen and a bog is in their water, which controls what plants *can* grow, and not the particular plants that *do* grow. For example, a bog doesn't have to contain sphagnum, although it usually does. What makes a bog a bog is its nutrient-deficient, acidic water, practically all derived directly from rain.

10.3 The Peculiar Behavior of Bogs

Once a bog is established, it sometimes expands beyond the borders of the depression in which it got started; it "overflows" its original basin. Sphagnum moss, the commonest plant in most bogs, holds huge quantities of water—it can absorb sixteen times its own weight.[9] Therefore, when sphagnum begins to grow over the edge of a bog, it takes the water it needs for growth with it; this allows it to continue growing outward indefinitely. In this way the original bog spreads, slowly converting dry ground to wetland. The process is called *paludification.*

Expansion sometimes continues until the bog covers a large area. The sphagnum grows upward as well as outward, and as it thickens, the bottom layer is gradually converted to peat, which lies like a blanket over the land. The land doesn't have to be flat: the bog can continue invading on gentle slopes, up, down, and across them, leaving peat draped over hills and hollows. When this happens, the bog is a *blanket bog.*

As a bog grows thicker, its center sometimes rises higher than its periphery, and what was a flat bog develops into a *domed* or *raised bog* (figure 10.2). The water table, which is the surface of the water held in the sphagnum and saturating it, rises inside the growing sphagnum, which means that the water

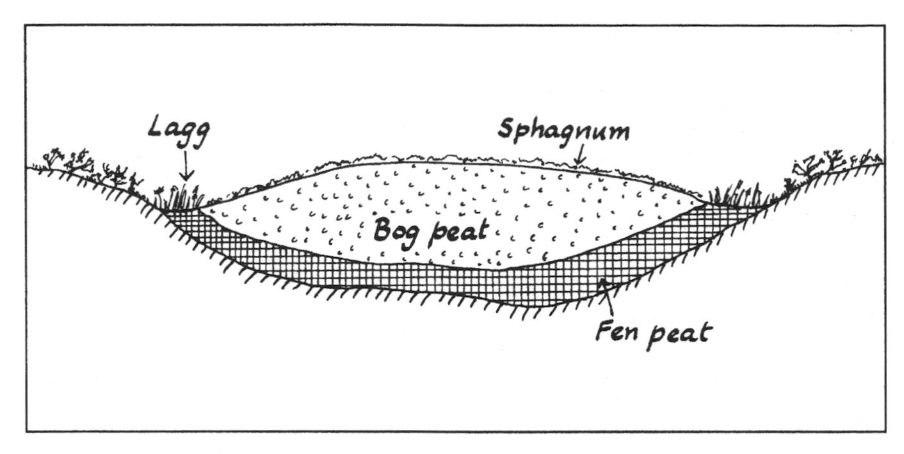

FIGURE 10.2. Vertical section through a domed bog. Sphagnum grows over the dome, and sedges grow in the lagg (moat) around it. Vertical scale exaggerated.

table itself becomes domed. Its highest point may be several meters above the water table in surrounding land, and the water drains outward from the bog center in all directions.

The mound of water filling a domed bog is not, as was long believed, held in place by capillary action, the force that holds water in a sponge.[10] In other words, the water mound does not amount to a capillary fringe atop a body of groundwater (see section 4.4 and figure 4.5). If it did, the water would be under tension and wouldn't seep in to fill the hole when a peat block is cut from the surface, and wouldn't drip from the block itself. In fact, the hole fills and the block drips, from which it follows that the water is not capillary water. It is simply collected rainwater.

The dome of a domed bog is a separate entity, in the hydrological sense, from the bog beneath it. The original bog on which the dome has grown usually receives at least some groundwater seepage; the wetland underlying the dome may even have been a fen, fed by a comparatively large proportion of nutrient-rich water. The water in the dome proper, however, is pure rain and nothing else. Thus a complete domed bog has two layers of peat.[11] The peat at the bottom is more like fen peat—it may even be fen peat—consisting of the remains of plants that require a modicum of nutrients and can endure only mildly acidic water. The peat at the top consists of the remains of "extreme" bog plants, those that can survive on nothing but nutrient-poor, strongly acidic rainwater; this kind of peat is called *ombrotrophic* (rain-fed) peat.

Some remnant of the richer wetland on which the dome developed usually persists around the margins of a domed bog, forming a moat (otherwise known as a *lagg*) where nutrient-requiring sedges and grasses can grow. As you walk—or wade—onto a domed bog, the vegetation you encounter shows how the groundwater is changing. At first, sedges and grasses indicate that the water is comparatively rich; as you continue, the ground level rises slightly, and sphagnum and other bog plants show that you have entered the strictly ombrotrophic part of the bog. Indeed, domed bogs are places where you can tell the origin of the groundwater at different spots merely by recognizing the plants growing there and knowing their ecological requirements.

Yet another kind of bog familiar to most people is the *quaking bog*. A quaking bog is a bog that has expanded over a pond, thickening as it spread until it became strong enough to support a person's weight. Walking on it is like walking on a waterbed; you sink in and may even lose your balance; you can see ripples spreading away across the moss surface at every footfall.

Bogs seem bleak and cheerless to many people, and the feeling isn't entirely subjective. A bog is, truly, a cold place to be. The reason is this:[12] peat freezes in winter, and ice-filled peat has a much greater thermal conductivity than either wet peat or ordinary soil. The outcome is that, while frozen peat is exposed to the air through late fall and early winter—before the first snowfall lays an insulating blanket over it—the ground's heat is lost more rapidly from a bog than from the surrounding dry land. The bog surface therefore becomes extra cold, and the low surface temperature leads to low ambient temperature, especially on cloudy days.

The southernmost outposts of the permafrost—the permanently frozen ground of the Arctic—are to be found in bogs, because they are always colder than the surrounding land. Permafrost has no exact southern limit; it simply becomes less and less widespread as you move south from arctic latitudes, and its southernmost occurrences are isolated patches in temperate-zone bogs.

10.4 Ribbed Fens and String Bogs

Some peatlands develop a most peculiar landform: they become corrugated. An array of long, narrow grooves alternating with long, narrow ridges develops. The grooves often fill with water. A wetland of this kind (figure 10.3) can be called a *ribbed mire* as a general term, a *ribbed fen* if it's a fen, or a *string bog* or *strangmoor* if it's a bog (as always with wetlands, there's no shortage of names).

The way in which these striking patterns develop is uncertain, although theories are numerous.[13] Fens and bogs can both become ribbed, and when it happens to a domed bog, the ridges and grooves make a pattern of concentric rings about the highest point.

The ridges—called *strings*—are always at right angles to the slope of the ground, and they act as dams blocking downslope drainage. Water collects behind them in the grooves as long, narrow, parallel pools, called *flarks*. The strings are spaced about 5 to 10 meters apart on average; because they lie across the slope—usually an exceedingly gentle slope—the flarks are at a sequence of different levels, forming a flight of shallow "stairs."

The fact that the strings function as dams shows that they must be impermeable, or almost impermeable, to water. It follows that the compressed peat forming them has very low hydraulic conductivity, orders of magnitude lower

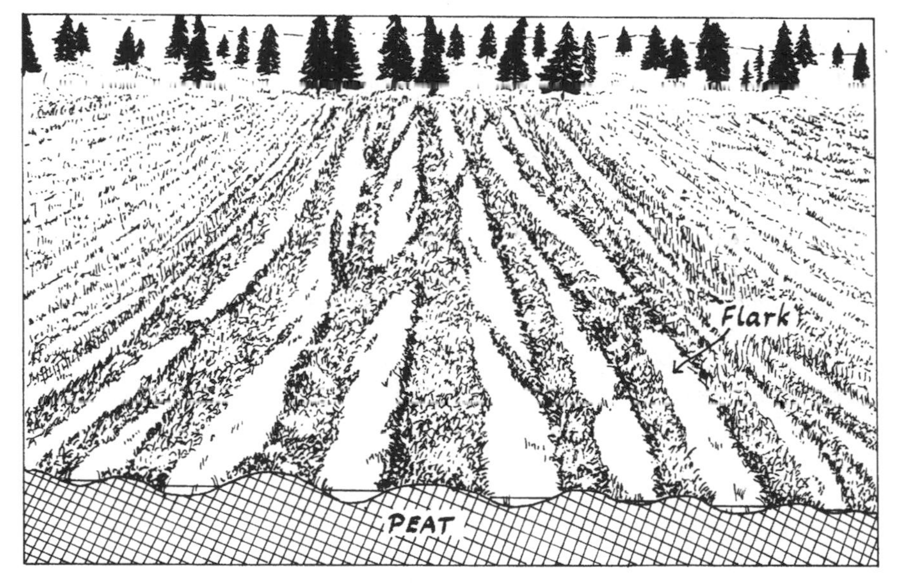

FIGURE 10.3. A ribbed mire. Note how the ponds (flarks) lie along the contours and descend, as shallow steps, from left to right.

than that of the uncompacted plant remains at the surface of the peatland. This explains why peatlands remain wet: the rainwater and overland flow that collects in them is retained as in a swimming pool, by an impermeable peat "liner." It also explains why peatlands are so different from wetlands that lack peat, which we consider next.

10.5 Other Wetlands: Swamps, Marshes, and Prairie Potholes

Peat cannot form—or, anyway, not a thick layer of it—in a seasonally flooded wetland, that is, one that dries out from time to time. Temporary dry periods allow decay-causing bacteria, which need oxygen, to attack dead, moist plants remains as soon as they are exposed to the air; the remains are then consumed by the bacteria—in other words, they decay—instead of accumulating as peat.

Wetlands lacking peat, or with only a thin layer of it, develop in warmer, drier climates than do peatlands. The thinness of the peat (when there is any) means that plants of nonpeaty wetlands are rooted, for the most part, in mineral soil. As a result their vegetation differs from peatland vegetation; it contrasts especially strongly with bog vegetation—bog plants are never found in a nonpeaty wetland.

Nonpeaty wetlands come in two forms: *swamps* and *marshes*. They resemble each other in being inundated in the growing season with still or slowly moving water that is richer in plant nutrients than bog water and much less acidic. The difference between them is in their vegetation. A nonpeaty wetland is defined as a swamp if it is treed, and as a marsh if it is covered with grasslike plants.

Swamps are hospitable to a wide variety of plants. The species depend on the location. As for the tree species, red maple, silver maple, and black ash are typical swamp trees in the midlatitude deciduous forests of eastern North America; redcedar dominates the swamps of the Pacific coast rainforest; water oak, bald cypress, and tupelo gum are common in the warm-climate swamps of the southeastern states.

Marshes are mostly covered with sedges, bulrushes, cattails, and reeds, sometimes accompanied by wetland shrubs such as willows (figure 10.4). The hydrological differences between a swamp and a marsh are not great. Treed swamps usually become established in places where, when the annual flooding subsides, the water table sinks to below a tree's rooting level.[14] Marsh plants don't require such well-aerated soil as trees, and can grow where the soil a

FIGURE 10.4. Some common marsh plants. *(a)* Baltic rush, *Juncus balticus; (b)* reed grass, *Glyceria grandis; (c)* cattail, *Typha latifolia; (d)* water sedge, *Carex aquatilis; (e)* great bulrush (tule), *Scirpus lacustris.*

short distance below the surface is constantly wet. Other factors besides the water on and in the ground affect the vegetation of a wetland, however, particularly its history, the fertility of its soil, and the vegetation on the dry land surrounding it.

Degrees of wetness form a continuum. Intermediate in wetness between marshes and dry land are *wet meadows,* which are sometimes classed as true wetlands, sometimes as semi-wetlands. They are inundated for shorter periods than are marshes, though their soil is always waterlogged to within a few centimeters of the surface. Meadows support a more varied vegetation than marshes; grasses, rushes, and a wider range of flowering plants replace typical marsh plants.

An outstanding characteristic of marshes is the way their vegetation grows in zones. Different plant species are adapted to being partly submerged for different lengths of time; those requiring almost constant wetness naturally grow in the deepest parts, and plants able to endure longer dry periods grow on higher ground. The elevation differences may seem negligible, but they control the pattern of the rising and falling water surface, and hence the length of time any given spot of ground is submerged. The match between plants and elevation is usually good enough for an air photo of a marsh to serve as a contour map of the ground surface; the boundaries separating the different vegetation zones mark the contours. Thus a marsh growing on gently sloping shores around a circular pond consists of concentric rings of contrasting vegetation.

An area where marshes are to be found in unbelievable abundance is the prairie pothole region of the central plains (figure 10.5).[15] The *potholes,* otherwise known as *sloughs* (rhymes with *blues*), are small ponds, each forming the center of a "miniature marsh"; it has been estimated that there are 4 or 5 million of them in "pothole country."[16] The individual ponds are small; their average diameter is probably about 50 meters. The vast number of small depressions required to contain all these ponds owe their existence to the great ice sheets of the last ice age, which disappeared from the region about 14,000 years ago, leaving behind a thick deposit of rubble (technically, *ground moraine*) pitted with innumerable hollows. Now, provided it hasn't been artificially drained, each hollow contains a pond.

The hollows vary in size and depth. Some are so small that they usually dry out completely every year, whereas others are large enough and deep enough to contain permanent ponds.[17] The water level is highest in spring, when each pond is topped up with water from melting snow, while groundwater seeps in because the water table is high. Thereafter, in spite of summer rains, the level drops as the season progresses. Much of the drop is caused by the marsh plants surrounding the pond, especially the willows: they suck up so much groundwater that the level of the water table falls, whereupon water seeps down out of the pond into the dry soil. Each pond is surrounded by its own concentric rings of marsh plants, which consist of the species adapted to that particular pond's depth and flooding schedule: the marsh around a shallow pond that dries up completely has a different set of plants than the marsh around a deep, permanent pond. Given the necessary botanical information, one can judge how the water behaves at a given place by merely observing the plants.

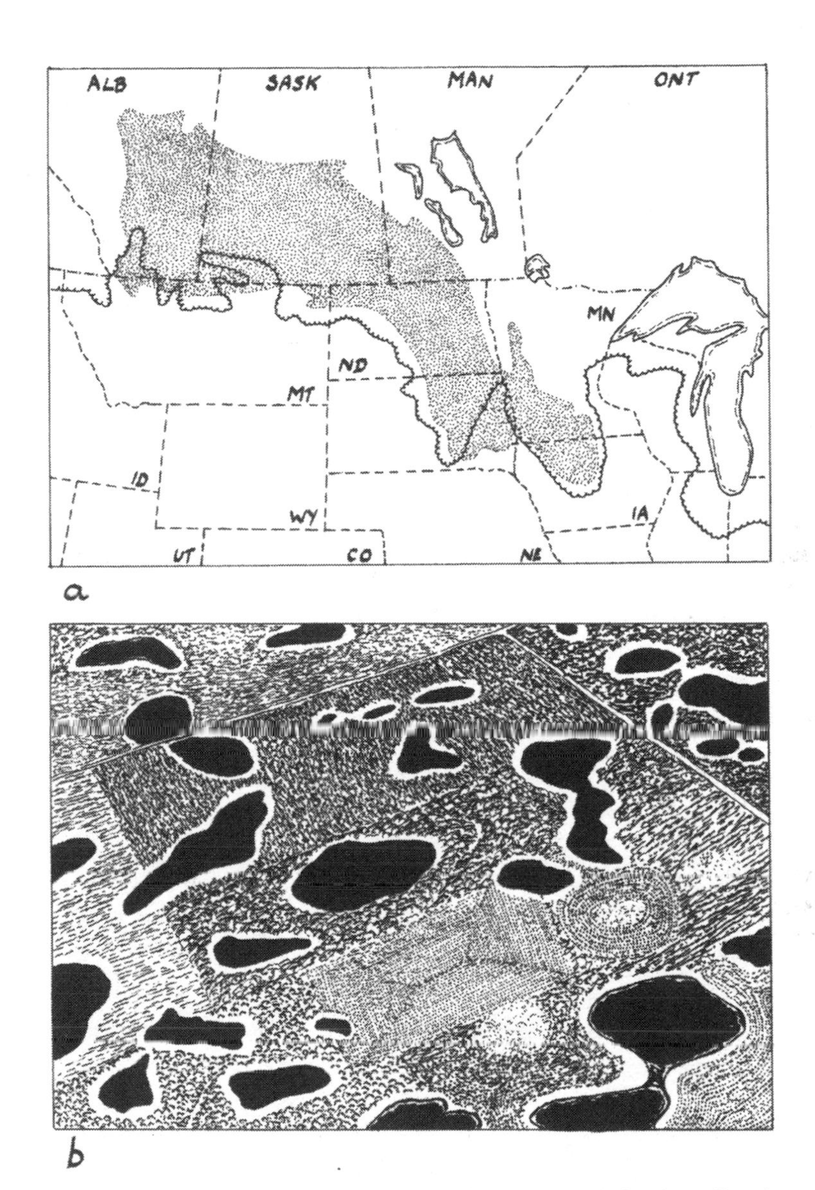

FIGURE 10.5. *(a)* Map showing prairie pothole region (stippled). The scalloped line shows, approximately, the southern boundary of the ice sheets of the last ice age 15,000 years ago. *(b)* Aerial view of prairie fields dotted with potholes.

The ponds with their surrounding marshes make the prairie pothole region one of the most important duck-nesting areas in North America. Filling in ponds to create additional arable land has reduced the numbers of ducks enormously. Luckily, the shrinking duck population has helped convince people of the folly of destroying or damaging natural wetlands.

11

Microscopic Life

11.1 The Different Kinds of Life

This book is about fresh water, so far with only incidental mention of the living things in it. In this chapter we consider the microscopic life in the water—the hosts of microscopic (or almost microscopic) creatures that are as much part of natural fresh water as peas are part of pea soup. The organisms are too small to be examined with the unaided eye, even those that are just visible as tiny specks; but they are abundant enough to affect the water in which they live, in a way that influences the welfare of all the larger animals and plants sharing it.

Before going further, we must look at how life as a whole is classified. Because of their small size,[1] it is tempting to lump all microscopic organisms as "microbes" and treat them as though they were indistinguishable. In fact, they are enormously diverse. To a biologist the contrast between a bacterium and an amoeba is far greater than the contrast between an amoeba and a human being.

Consider all living things.[2] All are made up of cells— sometimes only a single cell—and classification is based first on the structure of the cells and then on their genetic material.[3]

The first division is between the *Prokaryotes* and the *Eukaryotes.* Living cells exist in two fundamentally different forms, and those of the prokaryotes are much simpler and more primitive than those of the eukaryotes; a prokaryote cell has no nucleus, and its genetic material consists of a single DNA molecule. Eukaryote cells are much larger, and their DNA, together with other materials, is divided into a number of distinct chromosomes enclosed in a nucleus—a separate "package" inside the cell—walled off by its own membrane. Almost all the living plants and animals you will ever see consist of eukaryote cells. The most abundant prokaryotes are bacteria.

Prokaryotes are far more ancient than eukaryotes. The oldest traces of life on earth,[4] which date back more than 3.5 billion years, were left by prokaryotes. The earliest known eukaryotes didn't appear until about 1.4 billion years ago. It seems safe to say that the prokaryotes had the earth to themselves for over 2 billion years, whereas the eukaryotes, which presumably evolved from them, have been around for less than half that time.

Having divided all living things into these two great groups, the next tier of the classification is as follows:

PROKARYOTES		EUKARYOTES
Bacteria	*Archaea*	*Eukarya*

The second line shows the three great *domains* of the living world, each rich in diversity.

A dozen different evolutionary branches are recognized in the *Bacteria,* even though they are the smallest known organisms; some are no more than 0.5 micrometers long (1,000 micrometers make a millimeter). The disease-causing bacteria are the only "undesirables" in the bunch, and they are undesirable only from the human point of view. Bacteria live in a variety of environments. Those living in water are the ones we shall be concerned with here, but there are many others, for instance, soil bacteria and the gut-inhabiting bacteria without which most "higher" animals, including us, would be unable to digest food.

The *Archaea,* forming the second, possibly more ancient, domain, are another diverse group of prokaryotes. They have been put in a domain of their own for genetic reasons though they resemble the Bacteria in many respects. They differ in their ecology, however. They live in the most inhospitable environments on earth—water that seems, to us, excessively acidic, alkaline, salty, or hot: some thrive in hot springs at temperatures above 100°C. Most can live only where no air can reach them, as they would die if they were exposed to it. And some live in bogs where they generate methane, a subject we return to below.

The third domain, the *Eukarya,* embraces all the eukaryotes, and its evolutionary branches comprise all animals and plants, as well as the fungi, the slime molds, and a vast number of microscopic creatures collectively known as protists.[5] The protists are all either single-celled or made up of an aggregation—a "colony"—of similar, undifferentiated cells. Some contain chlorophyll, which colors them green and allows them to make their own food by photosynthesis; these are the organisms at the beginning of aquatic food chains. Others are nongreen, which puts them one or more links up the food chain. Independently of whether they are green or not, some have tiny whiplike appendages (*flagella*) enabling them to swim, and some don't. The nongreen protists used to be classed as animals and called Protozoa. The green ones were classed as plants and included with the Algae; they are still called *algae* colloquially, but we shall avoid the word here. Botanists and zoologists fought for proprietorship of the ones that are simultaneously animallike (in being able to swim actively) and plantlike (in being able to photosynthesize). But none are animals or plants in the ordinary sense, and it is more straightforward to treat them all as protists; in contrast to the protists, all other Eukarya are multicellular.

In spite of their larger size, the Eukarya are no more important than the Bacteria in the ecology of fresh waters, as we shall see in the following sections. The Archaea are comparatively unimportant because, owing to their aversion to oxygen, they occupy only a few unusual environments; even so, they may be having a marked effect on the outside world because the methane they produce is one of the greenhouse gases causing climatic warming.

In a nutshell, microscopic organisms in all their variety are vitally important to freshwater ecosystems. Millions of them live free in the water as part of the *plankton,* to be considered in the next section.

11.2 Lake Plankton: What It Consists Of

The plankton is the mass of living organisms suspended in a body of water and entirely surrounded by it; they are not in contact with any solid objects such as water plants, rocks, or mud, nor with the surface film.[6] In the strict (but not very useful) sense, *plankton* means passively floating organisms only, but in the ordinary sense it means these plus all the tiny creatures sharing the same space and able to swim feebly, though not vigorously enough to resist being carried this way and that by turbulence.

Plankton requires still rather than flowing water for its proper development. Flowing water sweeps the organisms away, preventing the buildup of big populations. Therefore plankton is typically found in lakes and ponds, not in rivers and streams.

The commonest plankton creatures are various bacteria and protists, together with some small animals, particularly rotifers and miniature crustaceans; these are true animals, notwithstanding their tiny size; unlike the unicellular protists, they are multicellular.

A few words on each of these groups, and see figure 11.1. "Small" bacteria—those less than 2 micrometers long—are abundant, and we postpone consideration of them to section 11.3. Lake plankton also contains plentiful "large" bacteria, however, with cells up to 60 micrometers long. These belong to the *Cyanobacteria,* also known as blue-green bacteria or blue-green algae[7] because of their color; besides a bluish pigment, they contain green chlorophyll and carry on photosynthesis. The chlorophyll is not confined to little "packages" (*chloroplasts*), as it is in most other green organisms; it is spread evenly through the interior of each cell. Some blue-green bacteria consist of solitary cells and some of colonies of cells joined together; they may be joined in chains (figure 11.1), sheets, or solid masses, but whether single or joined, the cells are nearly always imbedded in a clear jelly.[8] Blue-green bacteria require ample nutrient supplies, particularly nitrogen and phosphorus; they are apt to multiply extravagantly in eutrophic lakes in warm weather, and the result is a swarm of them, called a *bloom,* that makes the water cloudy and unattractive—"a teeming, green soup."[9]

Green protists[10] of many kinds are common in lake plankton. All contain chlorophyll, and although they are individually tiny, their sheer numbers ensure that their photosynthetic production outweighs that of all other water plants combined, in all but very shallow, weed-choked lakes.[11] Two common green protists in lake plankton, *Chlamydomonas* and *Volvox,* are shown in figure 11.1. Both can swim, weakly and slowly. A single chlamydomonas is invisibly small, but often they are present in such abundance they stain the water green. A volvox is a spherical colony of separate cells; although the individual cells are microscopically small, a whole colony is often big enough—about 2.5 millimeters across—to be just visible to the unaided eye as a tiny green ball rolling through the water.

The figure also shows a diatom, a member of another important group of

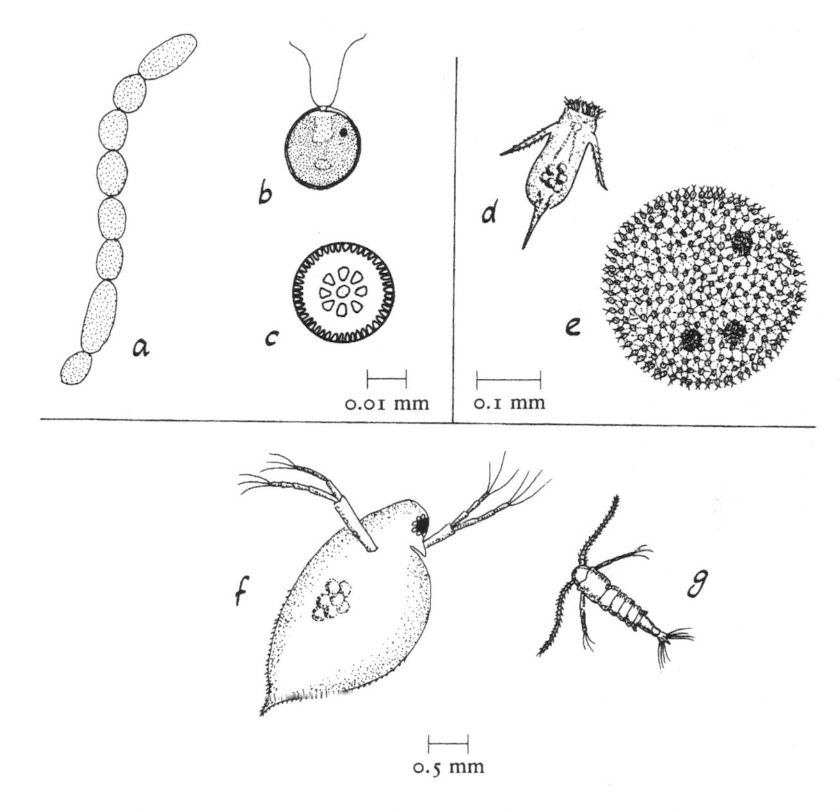

FIGURE 11.1. Some plankton organisms. Note the different scales of the three sections. *(a)* A cyanobacterium, *Anabaena; (b)* a green protist, *Chlamydomonas; (c)* top view of a "pillbox" diatom, *Cylotella; (d)* a rotifer, *Filinia; (e)* a colonial green protist, *Volvox; (f)* a water flea, *Daphnia; (g)* a copepod, *Diaptomus.*

protists. They are able to photosynthesize, but the green of their chlorophyll is masked by carotene (the pigment in carrots), which gives them a golden-brown color.[12] They are very distinctive protists. Each diatom is a cell encased in "walls" impregnated with silica. Its walls are in two parts, forming a shallow box with an overlapping lid, which resembles a pillbox if the diatom is circular or a shallow shoe box if it is more or less rectangular; there is a range of shapes—many diatoms are circular, others are oval, rectangular, diamond-shaped, dumbbell-shaped, or triangular—and the surfaces of their walls are often "engraved" with intricate and beautiful designs (it is a pity they are microscopic!). Diatoms are nearly always the dominant organisms in lake plankton,[13] and they are also abundant on solid surfaces, such as the lake bed, rocks, and water plants. The "grounded" ones are capable of moving; a diatom can glide along, driven by the movement of its protoplasm (the soft, living material forming the interior of the cell), some of which protrudes between the rigid "box" and the "lid" enclosing it. The ability to glide on a hard surface is no help to a diatom floating in the plankton, which is at the mercy of turbulence.

The plankton creatures we have considered so far are all at the beginning of the food chain—they are photosynthesizers that make their own food and then become the food of organisms higher up the chain, which can't photosynthesize. The simplest of these are many species of nongreen (chlorophyll-lacking) protists. But far more important are some small multicellular animals—the rotifers and two groups of tiny crustaceans, the water fleas and the copepods.[14]

Rotifers are the smallest multicellular animals; none are more than about half a millimeter long.[15] A rotifer's most interesting feature is its corona, a ring of short, constantly waving hairs (*cilia*) around its head. The way the cilia beat makes it look as though the whole corona is rotating: hence the name *rotifers* (wheel-bearers) and the old name *wheel animalcules*. The cilia propel the animal through the water and waft food into its mouth. The food consists of particles of dead organisms and bodies of live protists of many kinds; some large rotifers feed on smaller rotifers.[16] More than 1,500 different species of rotifers have been found in fresh water, but not all of them are part of the plankton: some live attached to plants.

Lastly, we come to the largest plankton animals: water fleas and copepods, most of them just big enough to be visible to the unaided eye. They are crustaceans, minute relatives of crabs and shrimps, which they resemble in having five pairs of legs.

The best-known water fleas are species of *Daphnia* (figure 11.1), which are familiar to anybody who has examined a sample of pond water with a dissecting microscope. A big one is 3 or 4 millimeters long, and it has a single large eye at the center of its "face" and big, hairy antennae extending over its head; the body is flattened laterally, like that of an ordinary flea; and the legs are concealed by a carapace that covers the body like a saddle. Water fleas swim energetically, using their antennae as oars. They are filter feeders: by waving their legs they set up currents that bring in floating food particles—

including protists—which are then strained out of the water by hairs on their legs.

The other minicrustaceans of the plankton, copepods, are not unlike water fleas, but differ from them in having cylindrical or pear-shaped, unflattened bodies, visibly divided into segments. Like water fleas they use their antennae as oars. The group as a whole have two feeding styles: some are filter feeders, like water fleas; others are carnivorous—they grab their prey in their mouths and chew it.[17] The filter feeders are the common kind in the plankton: the hunters mostly live on the bottom.[18] An example of the former kind, a species of *Diaptomus,* is shown in figure 11.1; it is about 2 millimeters long.

Water fleas and copepods are the strongest swimmers in the plankton; anything more energetic would be capable of choosing its own route through the water and its own destination, and would not count as plankton.[19]

11.3 Lake Plankton: How It Behaves

Plankton is passive but that doesn't mean it is unchanging. On the contrary, it undergoes seasonal changes as pronounced as those of terrestrial ecosystems on dry land. It burgeons in spring and dies down in the fall. The biggest changes are obvious to the most casual observer. They are the plankton blooms already mentioned. Because they are the only manifestations of plankton in action that most people ever see, they deserve further attention.

Conspicuous blooms consist of masses of blue-green bacteria (colloquially, blue-green algae or blue-greens), and they appear only in highly productive lakes; they are especially likely to appear in lakes polluted with sewage, which is always rich in phosphorus. The bacteria need warmth to multiply, so blooms occur mostly in late summer, after the hottest months have raised the water temperature and before it has had time to cool down.

A bloom can appear quite suddenly if masses of blue-greens produced by a subsurface population explosion float to the top. This is likely to happen if the deeper water becomes deoxygenated or otherwise unfavorable, whereupon tiny vesicles in the blue-greens' cells become filled with nitrogen, which makes them buoyant. Blooms can also be raised to the surface by currents; the circulating currents of Langmuir cells sometimes concentrate them in windrows (see section 7.5). They are also shifted by downwind currents: a bloom at the surface is often carried downwind to accumulate as a slimy layer on a lee shore.

Blue-greens are not the only plankton organisms that move up and down.

Many shift their positions throughout the day. The green, photosynthetic protists that are capable of swimming move up to profit from the daylight; they rise toward the surface in the morning and sink back down again in the afternoon or evening. Plankton animals do the opposite, and being much stronger swimmers and having no need for light, they can move faster and farther. Most of the water fleas and copepods make vertical migrations, but only a few of the rotifers. The fact that the green cells are moving up while the animals that feed on them are moving down, and vice versa, appears not to matter; the animals feed during their vertical journeys. The daily migrations are controlled by responses to light, which means they are much more pronounced in clear, transparent lakes, where the light penetrates deep, than in murky ones.

A few plankton animals are limited to particular water levels by their temperature requirements. Some can live only in the warm layer—the epilimnion—above the thermocline, and some only in the cold hypolimnion below it. So part of the plankton is "stratified," just as the water is. One of the water fleas that must have cold water, *Daphnia longiremis,* lives only in deep lakes in temperate latitudes but can thrive in shallow lakes at higher latitudes.[20] In temperate lakes it is restricted, during the summer growing season, to the cold, deep hypolimnion: it is trapped in the lakes' deepest layers. The hypolimnion of a deep lake and its living occupants are therefore at risk of being squeezed out of existence if global warming sends temperatures too high; animals higher up the food chain that depend on the animals of the hypolimnion as a food source would then starve.[21]

11.4 The Chemical Activities of Bacteria

Without bacteria, all other life on earth would soon grind to a halt. Most people are aware that all living things need nitrogen. Although supplies are ample—nitrogen as a gas forms 78 percent of the air—the pure form is not usable by the great majority of living organisms. Before plants can absorb nitrogen, it must be "fixed," that is, converted into simple inorganic compounds that plants can absorb, and nearly all the fixing is done by bacteria.[22]

In terrestrial ecosystems the fixing is done by bacteria living in the soil and, most importantly, by bacteria living inside the roots of certain plants, chiefly members of the bean family, which includes alfalfa and clover, much used by farmers to restore the fertility of their fields naturally.

In freshwater ecosystems the job is done by the blue-green bacteria—not by all species but by many of them. Thus the organisms responsible for unpleasant water blooms are vital to all other aquatic life. Moreover, in the evolutionary sense they have made all "higher" life (higher than other bacteria) possible. Blue-green bacteria have been around for over 3 billion years and are believed to have been the first organisms to carry on photosynthesis, beginning the conversion of the earth's original oxygen-lacking atmosphere to its present form.

The achievements of all other bacteria pale by comparison, but are still not negligible. Perhaps the next most important group—in the world in general, as well as in lakes and ponds—are those causing the decomposition (in other words, the decay) of dead animals and plants, and the recycling of all their materials. Many bacteria do this, some of them free-living, others as part of the bacterial flora living in the gut of scavengers of all kinds and mediating their digestive processes for them (digestion isn't a thing you do without the aid of bacteria). However, the bacteria busily decomposing organic remains are simultaneously breathing and therefore using up oxygen. More formally, they are creating a biochemical oxygen demand (BOD) in the water, as mentioned in section 7.8. The BOD of a sample of water can be measured and used as an indirect indicator of the level of pollution of lake water[23]—the bigger the population of bacteria, the larger the quantity of sewage or other waste there must be for them to consume. The bacteria doing the decomposition are not causing the pollution; they are cleaning it up.

Another important group of freshwater bacteria are those that cycle phosphorus, the plant nutrient most often in short supply. The phosphatizing bacteria were mentioned in section 7.7; they extract phosphorus from organic remains and convert it to phosphates usable by aquatic plants and blue-green bacteria. Too much phosphorus is worse than too little, however; pollution of a lake with added phosphates causes rampant blooms of blue-green bacteria, whose growth was previously checked by lack of phosphate. Nowadays, sewage pollution is the usual cause of excess phosphate; phosphate-containing detergents used to be a cause until they were banned. A lake is soon cleared of surplus phosphate once pollution stops; the cleanup is believed to be mostly the work of other bacteria, which consume the phosphate and carry it down into the lake sediment.[24]

Many other elements in lake water are "manipulated" by bacteria. We have already considered how bacteria interact with mercury (section 9.2). Next, consider sulfur. There are sulfate-reducing bacteria, which function only where oxygen is absent; they convert the sulfur in organic remains into hydrogen sulfide, the gas that gives rotten eggs their notorious stench. There are also sulfide-oxidizing bacteria, which convert sulfide into sulfuric acid. One of these, *Thiobacillus ferrooxidans,* "assists" in the production of acid mine drainage. What happens was described in section 7.8. The bacteria speed up the process enormously; without them it is slow. The sulfide affected is iron sulfide (pyrite), so the bacteria are affecting iron as well as sulfur.

Iron is attacked by several kinds of bacteria. Some iron-oxidizing bacteria live in bog ponds, where they convert soluble forms of iron (found mostly in the acidic water of bogs) into insoluble forms that settle out on the bottom, as a rust-colored precipitate.

Two colorful groups of freshwater bacteria proliferate abundantly in certain kinds of water. They are the purple and green sulfur bacteria; they use hydrogen sulfide, which is poisonous for most living organisms, as their energy source.[25] One of their chosen habitats is the clear, deoxygenated water at the bottom of strongly stratified lakes; hydrogen sulfide must be present and the water must be clear enough for daylight to penetrate down to the oxygen-free layer. The best lakes for these bacteria are permanently stratified (meromictic) lakes, where the blooms form thick mats on the bottom that sometimes persist all through the year. Unfortunately, the blooms are too deep to be seen from the surface, but they can be admired in shallow water in an entirely different habitat—hot sulfur springs. Some flourish at temperatures of 70°C.

Bacteria also live in groundwater; an assortment of different species has been found at a depth of 300 meters.[26] There is some hope that groundwater bacteria may be able to detoxify some of the toxic pollutants known to have found their way down into groundwater; the process is known as *bioremediation.* It is likely to be slow even if it happens, and shouldn't be relied upon as a solution to groundwater pollution problems; the need to prevent groundwater pollution is as great as ever.

11.5 Methane Producers

A weird, sinister light can sometimes be seen, on a calm, dark night, flickering on the surface of a bog (it happens more often in gothic fiction than in real life). It is known as a will-o'-the-wisp or an ignis fatuus—something that misleads or deludes. In fact, it is burning methane, a flammable gas consisting of carbon and hydrogen, given off from the waterlogged parts of wetlands and from lakes. And it is produced by the action of prokaryotes that, until recently, were classified as bacteria. The modern classification puts them in the Archaea, a domain separate from the Bacteria (section 11.1), but they are still usually called methane bacteria.

Methane bacteria can only live in mud or water wholly devoid of oxygen, and they can act only on a few particular carbon compounds produced by *other* bacteria; the process is far from straightforward, and it illustrates—not that you can actually see it—the complexity of the microscopic processes that underlie ecosystem functioning.

The methane produced by methane bacteria doesn't always reach the surface. As much as 90 percent of it[27] is used up by yet another group of bacteria—oxygen-requiring ones this time—that consume methane as *their* energy source. If a bog is the source of the methane, the methane-consuming bacteria live in the wet sphagnum moss at the bog surface; if the mud or deoxygenated water at the bottom of a lake is the source of the methane, the methane consumers live in the oxygenated water not far below the lake's surface.[28] In any case, not all the methane produced gets into the air, and of that which does, some, very occasionally, ignites spontaneously and burns as will-o'-the-wisps. An appreciable amount, however, is neither consumed nor burned: it becomes part of the atmosphere, where it acts as a greenhouse gas (see section 7.8); weight for weight, its effect is sixty times as great as that of carbon dioxide.[29]

Wetlands, lakes, and beaver ponds are all sources of methane. In the early 1990s it was estimated that 22 percent of the methane added to the atmosphere each year comes from natural wetlands.[30] This raises the question, how will global warming affect the methane flux? The question is, not surprisingly, unanswerable. Drought at midlatitudes may dry up much wetland, ending its methane output. At the same time the warmer climate will melt high-latitude permafrost, increasing the extent of active wetlands and causing methane production to go up. It is impossible to predict which of these effects will dominate; only time will tell.

Methane produced from natural lakes and wetlands is nowadays augmented by methane from artificially flooded land, as we saw in section 9.2. If the flooding produced by a dam inundates forested land, the "drowned" land plants, including the trees, provide plant remains for the methane bacteria (actually their precursors) to work on.[31] If the flooding inundates a bog, the bog produces more methane than it did before; this is because flooding drowns the methane-consuming bacteria that lived in the sphagnum layer at the top of the bog, leaving all the methane produced in the depths free to escape into the air; none is consumed on its way to the surface.

12

Water in the Atmosphere: Vapor, Clouds, Rain, and Snow

12.1 Closing the Cycle: Water Vapor

Previous chapters have discussed water, both liquid and frozen, on the land and underground. The time has come to close the water cycle by considering the water above the ground—some of it falling as rain and snow, some floating in the air as water droplets and ice crystals in the clouds, but most of it as invisible water vapor always and everywhere forming part of the atmosphere.

What comes down must go up. Obviously, the water cycle would come to a halt if the rain and snow arriving on the surface of land and sea, day after day all over the world, were not restored to the atmosphere—some of it after a temporary sojourn below the surface as groundwater. The passage from surface to atmosphere entails evaporation, the conversion of liquid water to water vapor. The molecules in liquid water are always in motion, and the warmer the water, the more rapidly they move. Molecules evaporate when their movement carries them up from the liquid surface to freedom in the air.

As liquid water changes to vapor, it becomes invisible. Although mist and the visible steam issuing from the spout of a kettle are often spoken of as "vapor," this is a misnomer; mist, steam, and fog consist of tiny liquid droplets. True water vapor is an invisible gas.[1] From an observer's point of view, there is therefore a tremendous contrast between the way water in its different forms is perceived. Water as a liquid or a solid (ice) can be seen. Water as a vapor can only be sensed by feel—and only vaguely at that—as a moistness, dampness, or mugginess in the air.

The amount of water vapor the air contains is known as its *absolute humidity;* it is measured as grams of water per cubic meter of air. The absolute humidity of the air cannot increase indefinitely; it cannot exceed an upper limit, known as the *saturation humidity,* which depends on the temperature: warm air can hold more vapor than cold. For example, at a temperature of −20°C, a cubic meter of air is saturated when it contains about 1 gram of water; at the freezing point (0°C), it can hold nearly 5 grams, and at 30°C more than 30 grams.[2]

The amount of vapor in unsaturated air is most conveniently measured by its *relative humidity* (RH); this is the absolute humidity as a percentage of the saturation humidity at the ambient temperature. For instance, air at 20°C is saturated by 17.3 grams of water per cubic meter; if it contains that much, its RH is 100 percent; if it contains half that, or 8.65 grams, its RH is 50 percent; and so on. The relative humidities encountered in nature range from almost (but never quite) zero, in a desert at midday, to 100 percent on a foggy day in a rain forest.

Another measure of the humidity is the *dewpoint.* If unsaturated air is cooled, its RH rises; when the RH reaches 100 percent, that is, when the vapor reaches saturation level, the temperature has been lowered to the dewpoint; thereupon, the vapor starts to condense as water droplets. When a glass of cold water fogs on the outside, it shows that the surface of the glass is at or below the dewpoint.

Water vapor rises from ice as well as from liquid water, in a form of evaporation known as *sublimation.* When a snowbank "wastes away" in winter sunshine, remaining crisp and dry all the while, it is sublimating.

12.2 Evaporation and the Water Cycle

Measuring the quantity of water vapor in the air at a particular place at a particular time is comparatively easy. Measuring the amount of water evaporating from a large tract of country per week or per month is considerably harder, but that is the figure needed by a hydrologist calculating the "water budget" of a region.

Water evaporates from the surfaces of lakes, ponds, and rivers. It also evaporates from the land, from moist soil, and, especially, from living vegetation. Vegetation pumps an enormous amount of water from the soil into the air; few people realize how much, because the whole process is invisible. For example, a single hectare of Douglas-fir forest spews out about 50 tons of water vapor in the course of a sunny, summer day, or about 235 bathtubs' full. Evaporation from the soil and transpiration from the vegetation combined is known as evapotranspiration, as we saw in chapter 4, and the transpiration part of it is much more important than the evaporation part, on most lands, in the growing season.

In measuring the total evapotranspiration from a region, as part of computing the region's water budget, water bodies and "dry" land are treated separately.

Estimating the evaporation rate from a lake or pond is easy in principle but (needless to say) less easy in practice. Except for an isolated lake with no inflows and outflows, evaporation accounts for only part of the lake's change in water level in dry weather. An obvious way to estimate the change due to evaporation alone is to set up a model water body—water in a pan—and measure the rate at which water evaporates from it.

Professional hydrologists use standard pans exposed on standard supports that allow the air to circulate over and around them.[3] A pan is left undisturbed for a chosen interval of time, and the drop in water level, measured in millimeters, is taken to represent the drop in level, caused by evaporation, that has taken place from a nearby lake or pond over the same interval; adjustments to allow for rainfall may be necessary. Because the pan water always warms up faster than lake water—and therefore evaporates faster—the drop in water level in the pan overestimates the matching drop in lake level, so the pan result is multiplied by a correction factor of about 0.75 (it varies slightly, from month to month).

To find the actual volume of water evaporated, it would be necessary to multiply the loss in depth by the lake's area. Often it's simpler just to leave the answer as a depth, which can then be directly compared with gains due to rainfall.

So far, so good, but there are pitfalls to guard against and corrections to make. Standard evaporation pans have vertical sides, and a nonstandard pan used as a substitute must also have vertical sides, so that the surface area of the water remains the same as the level goes down. The level must be topped up from time to time to prevent the depth of water in the pan from sinking too low, otherwise the water will warm up excessively. And the pan must be placed where pets, birds, and farm animals aren't tempted to treat it as drinking water.

A complication not so easily corrected for is the formation of a "vapor blanket" over a real lake.[4] As dry air moves out over a lake, it starts to pick up vapor the moment it crosses the shore; the humidity of the moving air continues to increase as it moves onward, and at the same time its evaporative power becomes less and less; if the lake is large, the air sometimes becomes saturated, so that evaporation stops before it reaches the downwind shore. Nothing like this happens over a pan used as a model lake, so the measured evaporation rate from a pan inevitably overestimates the rate from a real lake.

Nowadays, direct observation is replaced by computation: mathematical formulas have been devised that make it possible to compute fairly accurately the amount of water that has evaporated from a lake during a given time interval using data on averages (over the interval) of sunlight, wind speed, air temperature, water temperature, and dewpoint.

Now consider evapotranspiration from the land—evaporation from damp ground plus transpiration from all the plants growing on the ground. As with evaporation from a lake, you can either construct a model or use a formula. The model is a *lysimeter*, in effect an expensive flower pot filled with representative soil and plants and left outdoors, buried to the rim in soil, to "evapotranspire." Needless to say, innumerable details require attention, and the method is "both time-consuming and expensive";[5] also, it is useless for forests because of the size of the trees.

Fortunately, a formula (Thornthwaite's formula) has been devised[6] that allows *potential* evapotranspiration (PE) to be computed using nothing more than climatological data with adjustments for the latitude and the month; recall, from chapter 4, that PE is evapotranspiration when the water supply is unlimited. Additional data (and considerable faith!) are needed to infer the *actual* evapotranspiration (AE); in particular, one needs to know how much water the soil contains. In a groundwater discharge area, where soil water is constantly replenished, AE is usually close to PE; AE falls far short of PE in a groundwater recharge area, and to a very small amount where there is no soil to absorb rainwater. In an urban landscape, or over tracts of bare rock, AE is only about one-tenth of PE.

12.3 Water Aloft: Vapor, Clouds, and Fog

Many of the water molecules that become part of the water vapor in the atmosphere return whence they came quite soon. They are in constant motion, so some return to the water from which they evaporated, or to the ice from which they sublimated, without further ado. And some condense, as dew, onto any solid surface whose temperature falls below the dewpoint. If the dewpoint is below freezing, the vapor condenses[7] directly as ice and the result is *hoarfrost.* (Hoarfrost is not frozen dew; the vapor does not condense to liquid water on its way to becoming hoarfrost.)

The water vapor that remains in the air is the raw material for clouds or fog. When the air cools to the temperature at which its humidity reaches the saturation level, some of the vapor in it condenses as tiny droplets. If it cools at ground level, which happens when warm air moves over cold ground, the result is fog. If the warm air rises, cooling as it does so, condensation starts when its temperature falls to the dewpoint, and the droplets form clouds.

Except at very low temperatures—below −40°C—water vapor condenses into water droplets only when tiny particles are present to serve as nuclei on which condensation can take place. The *condensation nuclei* consist of such things as smoke particles, dust, crystals of sea salt, fly ash, and even free-floating bacteria. These are the tiny particles that, by scattering sunlight, make the sky and distant scenery blue in sunny weather; they are smaller than the particles that reduce visibility and produce haze. They are always present in abundance, and their presence is something to be thankful for. Without them the water vapor in the air would continue as vapor until the air was excessively supersaturated; then some tiny disturbance would cause sudden condensation followed by massive, devastating rainfalls.[8]

It's obvious, when you come to think of it, that vapor can condense on particles as small as those that make the sky blue: this is what is happening when puffy white cumulus clouds start to form in a clear blue sky. The droplets are small at first, and the clouds, when the sun shines on them, are brilliantly white.

As the air continues to rise, its temperature continues to fall; although its absolute humidity dwindles—because more and more vapor condenses to liquid—the saturation humidity decreases too, because of the falling temperature, so condensation continues. The droplets grow bigger, but not without limit: the largest droplets that can be produced by condensation acting alone have a diameter of about 0.05 millimeters. This is about the size to which the droplets forming cumulus clouds grow.

Cloud droplets are not stationary—and in moving they often collide. When they collide they coalesce, and this is one of the ways cloud droplets grow to the size of raindrops (another way is described below). Raindrops do not differ sharply from cloud droplets; one could say a droplet becomes a drop when it is heavy enough to fall from a cloud instead of continuing to float within it. It won't stay this size as it falls, however. A raindrop always evaporates to some extent on its way down; it needs to be more than 0.2 millimeters in diameter when it starts to fall if it is to reach the ground.[9] Indeed, much of the precipitation that falls from the clouds—raindrops, hailstones, and snowflakes—evaporates in midair. If it all evaporates before hitting the ground, the result is *virga,* a tattered curtain of falling rain or snow with clear sky underneath.

A cloud's appearance is strongly affected by the size and spacing of its droplets. A young (newly formed) cloud consists of a dense crowd of small droplets, and it looks pure white when the sun shines on it. As the cloud ages, however, many of its droplets coalesce, becoming bigger, fewer, and more widely spaced; by the time the cloud is old, it looks dark gray, even in bright sunshine, because it reflects light less efficiently than it did when young.[10] The notion that any cloud will appear white if the sun shines on it is a fallacy.

Liquid droplets are not the only ingredients of clouds. Ice crystals are often present too; cirrus clouds consist wholly of ice crystals. The crystals form in two ways: directly, by condensation of vapor into solid without the formation of liquid water; and indirectly, by the freezing of liquid cloud droplets.

The formless, white sheets of cirrostratus that often cover the sky at the end of a dry spell are ice clouds. When the sun shines through cirrostratus, you will usually see a *halo* around it, at about a hand span's distance from the sun itself (a halo is often unnoticed because its large radius means that it is well separated from the sun); a halo is caused by refraction of the sunlight as it passes through the ice crystals, which act as tiny prisms, and the presence of the halo, a narrow ring with the colors of a pallid rainbow, proves that the cloud consists of ice crystals.

Small water droplets don't freeze easily. They remain in liquid form, becoming strongly supercooled, as the temperature drops well below the freezing point. By the time the temperature in an ascending cloud has sunk to −20°C, however, a large proportion of its droplets will have frozen, and by −40°C all of them will have.

We now come to the other means (besides coalescence) by which cloud droplets grow. A cloud at a temperature between about −20°C and −40°C is a mixture of water droplets and ice crystals. The air's relative humidity is greater relative to ice than it is relative to water;[11] in other words, liquid water evaporates more readily than ice sublimates. Consequently, when the water droplets in a cloud give off water vapor (and shrink), the vapor produced condenses as ice onto neighboring ice crystals (which grow): that is, the ice crystals grow at the expense of the water droplets. This causes the growth of big ice crystals, which turn into big raindrops when they melt. In temperate latitudes heavy rain nearly always comes from clouds whose tops are freezing; lower, warmer clouds yield nothing much larger than drizzle; this is strong evidence that the spontaneous transfer of water from shrinking droplets to swelling ice crystals is what leads to big raindrops.[12] Raindrops vary in size from "drizzle drops" of about 0.2-millimeter diameter to hefty drops approaching 10 millimeters; in terms of weight, which gives a rainstorm its "feel," the heaviest drops weigh around 100,000 times as much as the lightest. That is why they strike the ground—and you—so forcefully.

It remains to consider snow. It comes from clouds made up wholly of ice crystals, when the weather is cold enough for the temperature to be below freezing right down to ground level. When water vapor condenses directly to ice, it forms delicate ice crystals with hexagonal symmetry. As is well known (and is possibly true, though unproven) no two ice crystals are the same; they are said to be as individual as fingerprints. Each crystal has an intricate pattern; it may be a perforated hexagon, or it may have six identical, diverging rays, decorated with a variety of spiky and leafy protrusions. Their convoluted shapes allow the crystals to become interlocked when they touch each other, and once several are inextricably joined, the result is a snowflake.

In cold weather the flakes reach the ground as snow. In mild weather they melt and become raindrops on the way down. Much of the rain that falls in temperate latitudes starts down from the clouds as snow and doesn't turn into rain until it is close to the ground.

12.4 Back to the Ground Again: Rain and Snow

After lingering in the atmosphere as vapor and cloud droplets, water resumes its travels around the water cycle when it returns to earth as rain or snow.

Rainwater (apart from acid rain) has an undeserved reputation for purity. All rain picks up a variety of contaminants, first as it forms and then as it falls. Many of the particles on which vapor condenses dissolve in the condensed droplets: this removes chemicals from the air by a process called *rain-out*. The raindrops then touch and dissolve additional particles as they fall: this is known as *wash-out*. Still more chemicals are dissolved when rain falls through the foliage of trees; the kinds of chemicals picked up depends on the species of tree.[13]

Although rain and snow are lumped together as precipitation in meteorological summaries, they play very different roles in the water cycle. Rainwater that falls on the land becomes surface water, soil water, or groundwater as soon as it reaches the ground. Snow remains unmelted on the surface, sometimes for months, taking no part in the water cycle; it simply accumulates, often in large quantities.

The greatest accumulations are in mountainous country, for two reasons. First is the obvious reason: mountaintops are colder than the surrounding lowlands, and the precipitation that lands on them is likely to come as snow at all seasons, and to remain on the ground without melting. The second reason is that in a cold climate, where the plains as well as the mountains receive their winter precipitation as snow, snow falls at the same places over and over again in the mountains, because of the way mountain slopes control the wind; in flat country, by contrast, snowfalls are scattered independently of each other over a much larger area.[14]

The snow is especially deep on maritime mountain ranges, such as the Cascades of Oregon and Washington and the Coast Mountains of British Columbia. The climate of these mountains is comparatively mild, and the precipitation tremendous. Winter brings frequent, big snowstorms, and in the warm intervals between them, the snow softens and becomes densely packed. As well, trickles of meltwater fill any remaining air spaces in the snow and refreeze as ice. The snow in the colder, drier climate of the Rockies, besides being less in quantity, behaves differently: because the temperature stays low, the snow remains light and fluffy. Sublimation—the conversion of snow into vapor, which can quickly dissipate—is much faster when snow is light and fluffy than when it is wet and dense.

In the lowlands, too, the contrasts between different geographic regions in the density of their snow is remarkable. Here are some data.[15] Snow densities at the end of winter, measured in kilograms per cubic meter, were 430 in the British Columbia Coast Mountains, where snow is wet and dense; 330 on the open arctic tundra where snow is packed hard by constant, strong winds; and 260 in the much more protected boreal forest or taiga (note that a cubic meter of water weighs 1,000 kilograms, or 1 metric ton).

This brings us to the topic of snow as a natural reservoir. The water content of the snowpack is a matter of vital concern to people living in mountainous country. The pack is repeatedly inspected, monitored, and worried over, all through winter and into spring. The area covered by snow in a city's watershed is estimated from air photos. Snow depth is often determined by sampling the snow at numerous sites with a long boring tube[16] and weighing the core of snow retrieved to discover its water content. Another method is to record the weight of the snow that has collected at a number of representative sites using *snow pillows;*[17] a snow pillow is a flexible fluid-filled bag with a device attached for measuring the pressure on it, and one is placed on the ground at each of the sample sites before the year's first snowfall. Whatever method is used to estimate the volume of snow in an area and its water equivalent, difficulties arise because of the way snow drifts; this makes it hard to judge whether the sampling sites add up to a fair representation of the area being sampled.

To people who depend on winter snowpack for much of their water supply, the rate at which the pack melts is as important as the amount of water it yields. The deep snowpack on the mountains in a wet, coastal climate sometimes melts too fast; it is apt to happen when the snowpack is "warm" (that is, with the temperature at freezing point all through) and "ripe" (with pores already filled with water and unable to hold any more). Then a *rain-on-snow* event[18] will likely cause flooding: warm, humid air comes in from the ocean, heavy rain falls, and the snow melts with a rush.

Surprisingly, the warm rain is not the source of the warmth that causes the sudden melt. Rather, it is the heat liberated when vapor in moist air condenses on the snow surface. Condensing vapor gives off heat; this is the opposite of what happens when water evaporates from a wet surface—a sweating body, for example—in which case the surface is cooled because heat is absorbed from it. Further, the moist air accompanying rain-on-snow is nearly always kept moving by turbulence; this means that the air in contact with the snow is constantly renewed, and condensation continues unceasingly. In addition, radiant heat radiates down from the cloud base. Lastly, the rain itself imparts some warmth, though this is the least important heat source. When all these causes act together, and the ground itself is still frozen so that water can't soak into it, the meltwater pours down the mountainsides in floods. The floods do damage, and the stored water is lost before it can be used.

Catastrophic snow melts don't happen everywhere. In dry climates, snow is as likely to sublimate as to melt. Much of it simply disappears in the spring sun, leaving no moisture behind.

Vapor rising from snow is water becoming invisible again. But it is still part of the water cycle, and can never leave it. The same is true of the invisible steam coming from a teakettle's spout, or from the mouth of anything that breathes. Water in one form or another circulates around us, all the time and everywhere.

Notes

CHAPTER ONE

1. J. Martinec, 1985, Time in hydrology, in *Facets of hydrology II,* ed. J. C. Rodda, 249–90 (New York: John Wiley & Sons).

2. S. L. Postel, G. C. Daily, and P. R. Ehrlich, 1996, Human appropriation of renewable fresh water, *Science* 271:785–88.

3. The negligible changes are losses caused by diffusion of water vapor into space from the top of the atmosphere, and gains caused by emission of water vapor in volcanic eruptions. See, for example, C. R. Longwell, R. F. Flint, and J. E. Sanders, 1969, *Physical geology* (New York: John Wiley & Sons).

4. L. A. Frank, 1990, *The big splash* (Secaucus, New Jersey: Carol Publishing).

5. *Tons* denotes metric tons; 1 ton = 1,000 kilograms.

6. Postel, Daily, and Ehrlich, Human appropriation. Also, P. H. Gleick, ed., 1993, *Water in crisis: A guide to the world's fresh water resources* (New York: Oxford University Press).

7. Martinec, Time in hydrology.

CHAPTER TWO

1. J. Martinec, 1985, Time in hydrology, in *Facets of hydrology II,* ed. J. C. Rodda, 249–90 (New York: John Wiley & Sons).

2. C. W. Fetter, 1988, *Applied hydrogeology* (New York: Macmillan).

3. The data are from R. Brassington, 1988, *Field hydrology* (New York: Halsted Press).

4. Fetter, *Applied hydrogeology.*

5. A measure of a material's permeability that depends only on the characteristics of the material is its *Intrinsic permeability.* It is proportional to *Hydraulic conductivity* × *Viscosity/Density.*

6. This equation is known as Darcy's law. Readers familiar with DC electrical circuits will notice its resemblance to the equation *Current (amperes)* = *Voltage (volts)* × *Conductance (mhos).*

7. For a full description of the method, see R. S. Bradley, 1985, *Quaternary paleoclimatology* (Boston: Allen & Unwin).

8. P. Fritz and J. C. Fontes, 1980, *Handbook of environmental isotope geochemistry* (Amsterdam: Elsevier).

9. Fetter, *Applied hydrogeology.*

10. J. A. Cherry, 1987, Groundwater occurrence and contamination in Canada, in *Canadian aquatic resources,* ed. M. C. Healey and R. R. Wallace, 387–426 (Ottawa: Department of Fisheries and Oceans).

11. A confining layer is sometimes called an *aquiclude* and a leaky confining layer an *aquitard,* because they exclude and retard water, respectively.

12. To a physicist the word *piezometer* means a delicate instrument used in the laboratory, but a piezometer used outdoors is a sturdy device for measuring the pressure (or its substitute, the head) of groundwater.

13. A piezometric surface is sometimes called a *potentiometric surface.*

14. If the aquifer material is homogeneous enough for its hydraulic conductivity to be the same throughout, then *Transmissivity* = *Hydraulic conductivity* × *Thickness of aquifer,* and the equation in the text could be rewritten as *Volume per second* = *Hydraulic conductivity* × *Cross-sectional area of aquifer* × *Hydraulic gradient.*

15. Fetter, *Applied hydrogeology.*

16. The rate, which is known as the *geothermal gradient,* varies from place to place. Ordinarily it ranges between 1°C per 100 meters (see Fetter, *Applied hydrogeology*) and 1°C per 30 meters (see C. R. Longwell, R. F. Flint, and J. E. Sanders, 1969, *Physical geology* [New York: John Wiley & Sons]).

17. P. Meyboom, 1967, Mass-transfer studies to determine the groundwater regime of permanent lakes in hummocky moraine of western Canada, *Journal of Hydrology* 5:117–42.

18. D. R. Lee, 1977, A device for measuring seepage flux in lakes and estuaries, *Limnology and Oceanography* 22:140–47.

19. S. K. Frape and R. J. Patterson, 1981, Chemistry of interstitial water and bottom sediments in Perch Lake, Chalk River, Ontario, *Limnology and Oceanography* 26:500–517.

20. T. C. Winter, 1981, Effects of water-table configuration on seepage through

lakebeds, *Limnology and Oceanography* 26:925–34. Also, Fetter, *Applied hydrogeology.*

21. Meyboom, Mass-transfer studies.

22. This change in direction is the *refraction* of flow lines. It results from an abrupt change in the speed of flow; in exactly the same way, the change in the speed of light as it passes from, say, air into water causes the refraction (bending) of light.

23. Cherry, Groundwater occurrence.

24. Flow lines cross equipotential lines at right angles in isotropic aquifers, but not in anisotropic ones. This doesn't show in diagrams; the difference between the vertical and horizontal scales (done to ensure clarity) distorts the angles.

CHAPTER THREE

1. P. R. Ehrlich, A. H. Ehrlich, and J. P. Holdren, 1977, *Ecoscience: Population, resources, environment* (San Francisco: W. H. Freeman & Co.).

2. J. A. Cherry, 1987, Groundwater occurrence and contamination in Canada, in *Canadian aquatic resources,* ed. M. C. Healey and R. R. Wallace, 387–426 (Ottawa: Department of Fisheries and Oceans).

3. Figures 3.1a and 3.1b, respectively, are adapted from figures in C. W. Fetter, 1988, *Applied hydrogeology* (New York: Macmillan); and British Columbia Ministry of Environment, Lands, and Parks, 1994, *Groundwater resources of British Columbia* (Victoria: BC Environment; Ottawa: Environment Canada).

4. Cherry, Groundwater occurrence.

5. G. Fortin, G. van der Kamp, and J. A. Cherry, 1991, Hydrogeology and hydrochemistry of an aquifer-aquitard system within glacial deposits, Saskatchewan, Canada, *Journal of Hydrology* 126:265–92.

6. The older and younger tills were laid down, respectively, by the Illinoian and Wisconsinan glaciations. See R. J. Fulton, ed., 1984, *Quaternary stratigraphy of Canada,* Geological Survey of Canada, Paper 84-10 (Ottawa: Canadian Government Publishing Centre).

7. Fetter, *Applied hydrogeology,* 351.

8. Cherry, Groundwater occurrence.

9. Ibid.

10. W. B. White, D. C. Culver, J. S. Herman, T. C. Kane, and J. E. Mylroie, 1995, Karst lands, *American Scientist* 83:450–59.

11. Fetter, *Applied hydrogeology.*

12. R. C. Bocking, 1987, Canadian water: A commodity for export? in *Canadian aquatic resources,* ed. M. C. Healey and R. R. Wallace, 105–35 (Ottawa: Department of Fisheries and Oceans).

13. Electrical tests sometimes entail passing a direct current through the ground between buried electrodes, to measure DC electrical resistivity, and sometimes entail

sending electromagnetic waves into the ground from coils held above the surface, to measure electromagnetic conductivity. See J. J. Sweeney, 1984, Comparison of electrical resistivity methods for investigation of ground water conditions at a landfill site, *Groundwater Monitoring Review* 3:47–59.

14. Fetter, *Applied hydrogeology.*

15. To facilitate comparisons, all withdrawal rates given here are in the same units, namely, liters per minute, a unit commonly used in the context of household wells. Supplies to big cities are usually given in terms of cubic meters per day or per second. To convert, note that 1 cubic meter equals 1,000 liters. Also, since 1 cubic meter of water weighs 1 metric ton, it is briefer to speak of tons than of cubic meters.

16. Fetter, *Applied hydrogeology.*

17. R. Brassington, 1988, *Field hydrology* (New York: Halsted Press). The quantity used by a family of five in the United States is said to be about 2.3 liters per minute (E. Zwingle, 1993, Wellspring of the High Plains, *National Geographic Magazine,* March).

18. The figure is adapted from Brassington, *Field hydrology.*

19. The figure is based on data in C. W. Fetter, Jr., 1981, Interstate conflict over ground water: Wisconsin-Illinois, *Ground Water* 19:201–13.

20. Fetter, *Applied hydrogeology,* 278.

21. Bocking, Canadian water.

22. The term *carbonate* is used here, for simplicity, to include both carbonate and bicarbonate ions.

23. Cherry, Groundwater occurrence.

24. Ehrlich, Ehrlich, and Holdren, *Ecoscience.*

25. Fortin, van der Kamp, and Cherry, Aquifer-aquitard system.

26. I. J. Winograd and F. N. Robertson, 1982, Deep oxygenated ground water: Anomaly or common occurrence? *Science* 216:1227–30.

27. Fetter, *Applied hydrogeology.*

28. A detailed account of groundwater pollution and how it is spread is found in Cherry, Groundwater occurrence.

29. Ibid.

30. Ibid., 401.

CHAPTER FOUR

1. No distinction is made here (but see chapter 10) between pure peat, consisting entirely of organic matter, and so-called mineral soils, made up of a mixture of mineral and organic matter.

2. Some observed interception percentages in forests and grasslands are given on pages 187 and 220, respectively, in *Canadian aquatic resources,* ed. M. C. Healey and R. R. Wallace, 1987 (Ottawa: Department of Fisheries and Oceans).

3. *Overland flow* is occasionally, and confusingly, described as *runoff. Runoff* normally means *all* the water flowing out of a drainage area, only a fraction of which is overland flow.

4. W. G. Whitford, 1992, Effects of climatic change on soil biotic communities and soil processes, in *Global warming and biological diversity,* ed. R. L. Peters and T. E. Lovejoy, chapter 9 (New Haven: Yale University Press).

5. R. Buchsbaum and L. J. Milne, 1967, *The lower animals* (Garden City, New York: Doubleday & Co.), 133.

6. National Wetlands Working Group, 1988, *Wetlands of Canada,* Ecological Land Classification Series, no. 24 (Ottawa: Canadian Government Publishing Centre, Supply and Services), 79.

7. C. W. Fetter, 1988, *Applied hydrogeology* (New York: Macmillan).

8. Fetter, *Applied hydrogeology.*

9. Tensiometers sold as moisture probes usually measure *suction,* equivalently, *tension* (negative pressure); they give a positive measurement when the suction cup is in unsaturated soil. The number represents the force required for plants to remove water from the soil.

10. E. C. Pielou, 1988, *The world of northern evergreens* (Ithaca: Cornell University Press), gives an elementary account of how water moves through trees and other plants.

11. A small proportion of the water retained by plants is chemically combined with other elements (principally carbon) to construct new plant material, such as sugars and starches. Most of the water remains as water, and gives rigidity to the plant's cells.

12. T. A. Black, C. S. Tan, and J. U. Nnyamah, 1980, Transpiration rate of Douglas-fir trees in thinned and unthinned stands, *Canadian Journal of Soil Science* 60:625–31.

13. J. H. Richards and M. M. Caldwell, 1987, Hydraulic lift: Substantial nocturnal water transport between soil layers by *Artemisia tridentata* roots, *Oecologia* 73:486–89; T. E. Dawson, 1993, Hydraulic lift and water use by plants: Implications for water balance, performance, and plant-plant interactions, *Oecologia* 95:565–74.

14. T. E. Dawson and J. R. Ehleringer, 1991, Streamside trees that do not use stream water, *Nature* 350:335–37.

15. An ordinary hydrogen atom has a single proton as its nucleus. The nucleus of a deuterium ("heavy hydrogen") atom consists of a proton and a neutron, and therefore weighs twice as much. A molecule of "normal" water (H_2O) consists of an oxygen atom linked to two "normal" hydrogen atoms; in a molecule of "heavy" water (HDO), one of the normal hydrogen atoms is replaced by an atom of deuterium.

16. Field capacity is also measured as the proportion of a soil sample that consists of water when the soil contains all the capillary water it can hold; thus it resembles *specific retention* (see section 2.2). But when specific retention is measured, care is

taken to allow time for *all* the gravitational water to drain from the sample before it is weighed; this is not done when field capacity is measured.

17. The formula most often used is that of Thornthwaite and Mather. See C. W. Thornthwaite and J. R. Mather, 1957, *Instructions and tables for computing potential evapotranspiration and the water balance,* Publications in Climatology, vol. 10, no. 3 (Centerton, New Jersey: Drexel Institute of Technology). The formula is given in many publications on hydrology.

18. *Surplus* is here defined as *all* the precipitation not used in evapotranspiration. It is sometimes defined differently, to mean only the overland flow (or *overflow*).

19. The data are from *Groundwater mapping and assessment in British Columbia,* vol. 2, *Criteria and guidelines,* 1993 (Victoria: Resource Inventory Committee, Ministry of Environment, Lands, and Parks).

20. N. L. Stephenson, 1990, Climate control of vegetation distribution: The role of water balance, *American Naturalist* 135:649–70.

CHAPTER FIVE

1. C. H. Crickmay, 1974, *The work of the river* (London: Macmillan).

2. Gaining and losing streams are sometimes called, respectively, *effluent* and *influent* streams. Note that the prefixes (*e-*, out of, and *in-*, into) relate to the groundwater's "point of view." Groundwater flows *out* to an effluent stream. It is recharged by water flowing *in* from an influent stream.

3. National Topographic System of Canada, sheet 92/F5 (Bedwell River 1980), scale 1:50,000.

4. Although drainage density is a ratio, it is not a dimensionless number. Therefore, its numerical value depends on the units used for measuring length and area. For example, a drainage density of 12 kilometers of channel per square kilometer is equal to $0.6 \times 12 = 7.2$ miles of channel per square mile, since 1 kilometer = 0.6 miles.

5. A. N. Strahler, 1960, *Physical geography,* 2d edition (New York: John Wiley & Sons).

6. The groundwater flow from a topographically defined watershed may occasionally gain some water from, or lose it to, adjacent watersheds; this happens when groundwater divides fail to coincide with topographic divides (see section 2.7).

7. L. B. Leopold, 1994, *A view of the river* (Cambridge: Harvard University Press).

8. A. H. Laycock, 1987, The amount of Canadian water and its distribution, in *Canadian aquatic resources,* ed. M. C. Healey and R. R. Wallace, 13–42 (Ottawa: Department of Fisheries and Oceans).

9. See N. D. Gordon, T. A. McMahon, and B. L. Finlayson, 1992, *Stream hydrology: An introduction for ecologists* (New York: John Wiley & Sons). Leopold, *View of the river,* recommends a correction factor of 0.8.

10. Gordon, McMahon, and Finlayson, *Stream hydrology.*

11. J. B. Stewart and J. W. Finch, 1993, Application of remote sensing to forest hydrology, *Journal of Hydrology* 150:701–16.

12. Gordon, McMahon, and Finlayson, *Stream hydrology.*

13. More precisely, lag time should be measured from the moment at which 0.5 of the storm's total rain has fallen, and to determine the moment precisely, a continuously recording rain gauge must be used. Measuring time from the halfway point of the rain's duration, as in the figure, usually gives a close approximation.

14. Leopold, *View of the river.*

15. In Chézy's equation, *V,* the velocity, is proportional to $(RS)^{1/2}$. In Manning's equation, *V* is proportional to $R^{2/3}S^{1/2}$. Here *R* is the *Hydraulic radius = Cross-sectional area/Wetted perimeter; S* is the slope of the riverbed or the water surface, which are assumed to be equal.

16. Gordon, McMahon, and Finlayson, *Stream hydrology.*

17. Ibid.

18. H. B. N. Hynes, 1970, *The ecology of running waters* (Toronto: University of Toronto Press).

19. Gordon, McMahon, and Finlayson, *Stream hydrology.*

20. The volume of stormwater is given by the area under the curve; this is the same in all three hydrographs in figure 5.5.

21. Crickmay, *Work of the river;* Leopold, *View of the river.*

22. Leopold, *View of the river.*

23. This assumes that a change in depth is always accompanied by a change in cross-sectional area, which is always true in practice though it is not a logical necessity.

24. The principle and its consequences were first stated formally by Leonardo da Vinci in 1500, but its truth must have been recognized informally since a much earlier time.

25. Strictly speaking, waves "feel the bottom" (i.e., are slowed by it) only when their wavelength is less than one-fourth of the depth of the water. See, for example, R. E. Thomson, 1981, *Oceanography of the British Columbia coast* (Ottawa: Department of Fisheries and Oceans).

26. Hynes, *Ecology of running waters.*

27. D. Bershader, 1990, Fluid physics, in *Encyclopedia of physics,* 2d edition, ed. R. G. Lerner and G. L. Trigg (New York: VCH Publishers).

28. G. Petts and I. Foster, 1985, *Rivers and landscape* (London: Edward Arnold).

29. S. Vogel, 1981, *Life in moving fluids* (Princeton: Princeton University Press).

30. The viscous (laminar) layer is more correctly described as a *sublayer* because it is part of the *boundary layer* at the bottom of a stream where the velocity of flow is less than the free stream velocity. The boundary layer has two sublayers: a viscous sublayer below and a turbulent sublayer above.

31. I. R. Smith, 1975, *Turbulence in lakes and rivers,* Scientific Publication no. 29,

Freshwater Biological Association, UK, quoted in Gordon, McMahon, and Finlayson, *Stream hydrology.*

32. For mathematical accounts of unsteady laminar flow and vortex shedding, see Gordon, McMahon, and Finlayson, *Stream hydrology;* and Vogel, *Life in moving fluids.*

CHAPTER SIX

1. C. H. Crickmay, 1974, *The work of the river* (London: Macmillan).

2. I write "moving fluids" rather than "flowing water" so as not to exclude from the definition the tremendous erosive power of wind-borne sand, a more important cause of erosion in arid regions than water-borne sediment.

3. The speeds are calculated from the formula $V = 0.155\sqrt{d}$, where V is the current speed in meters per second, and d is the diameter of the rock in millimeters. See N. D. Gordon, T. A. McMahon, and B. L. Finlayson, 1992, *Stream hydrology: An introduction for ecologists* (New York: John Wiley & Sons), 331.

4. The suspended load is sometimes divided into *suspended bedload,* consisting of material picked up from the bed, and *washload,* material washed into a river from the land that has never settled on the bottom and may never do so.

5. J. Whittow, 1984, *The Penguin dictionary of physical geography* (New York: Penguin Books).

6. Sediment concentration is sometimes given in milligrams per liter (mg/l), which is roughly equal to parts per million (ppm), since 1 liter of pure water weighs 1 kilogram, and 1 kilogram = 1,000,000 milligrams.

7. The numbers are derived from sediment rating curves (plots of sediment discharge versus water discharge) given for the Tanana River in L. B. Leopold, 1994, *A view of the river* (Cambridge: Harvard University Press), figure 11.9; and for the Snake and Clearwater Rivers in Gordon, McMahon, and Finlayson, *Stream hydrology,* figure 7.22.

8. Crickmay, *Work of the river.*

9. J. N. Holeman, 1968, The sediment yield of major rivers of the world, *Water Resources Research* 4:737–47, quoted in Gordon, McMahon, and Finlayson, *Stream hydrology.*

10. The atlas gives the volume as that of a cone of height 300 meters and basal angle of 45°. The exact length of the sides of a cube of the same volume is 304.6 meters.

11. P. A. Carling, 1992, In-stream hydraulics and sediment transport, in *The rivers handbook,* ed. P. Calow and G. E. Petts, 101–25 (Oxford: Blackwell Scientific).

12. The technical difference between ripples and dunes is that ripples form on a hydraulically smooth bed and dunes on a hydraulically rough bed; see section 5.9.

13. Whittow, *Dictionary of physical geography.*

14. Crickmay, *Work of the river.*

15. Ibid.

16. The maps were traced from 1:250,000 maps of the National Topographic System of Canada, sheets 117A (Blow River 1989) and 117B (Davidson Mountains 1985).

17. Leopold, *View of the river.*

18. As water flows down a slope, it gains kinetic energy, K, and loses potential energy, P. The difference, $K - P$, changes continuously. In a meandering river, the magnitude of $K - P$ at any moment depends on the way energy is being dissipated on various "tasks," among them dislodging material from the banks and redirecting the flow. The *action* performed by the water in a given time interval is defined as the sum of all the instantaneous differences, $K - P$, during the interval. (Note that in physics the word *action* has this exact meaning, with no vagueness.) The principle of least action states that these instantaneous differences will change continuously in such a way as to minimize the action. See A. Watson, 1986, Physics: Where the action is, *New Scientist,* January, 42–44.

19. R. I. Ferguson, 1973, Sinuosity of supraglacial streams, *Geological Society of America Bulletin* 84:251–56.

20. C. R. Longwell, R. F. Flint, and J. E. Sanders, 1969, *Physical geology* (New York: John Wiley & Sons).

21. P. Jones, 1983, *Hydrology* (Oxford: Basil Blackwell).

22. E. A. Keller and W. N. Melhorn, 1978, Rhythmic spacing and origin of pools and riffles, *Geological Society of America Bulletin* 89:723–30.

23. Leopold, *View of the river;* H. B. N. Hynes, 1970, *The ecology of running waters* (Toronto: University of Toronto Press).

24. R. L. Beschta and W. S. Platts, 1986, Morphological features of small streams: Significance and function, *Water Resources Bulletin* 22:369–79.

25. M. Church, 1992, Channel morphology and typology, in *The rivers handbook,* ed. P. Calow and G. E. Petts, 126–43 (Oxford: Blackwell Scientific).

26. G. E. Grant, F. J. Swanson, and M. G. Wolman, 1990, Pattern and origin of stepped-bed morphology in high gradient streams, western Cascades, Oregon, *Geological Society of America Bulletin* 102:340–52.

27. Whittow, *Dictionary of physical geography.*

28. Hynes, *Ecology of running waters:* H. B. N. Hynes, 1983, Groundwater and stream ecology, *Hydrobiologia* 100:93–99; J. A. Stanford and J. V. Ward, 1988, The hyporheic habitat of river ecosystems, *Nature* 335:64–66; J. A. Stanford and J. V. Ward, 1993, An ecosystem perspective of alluvial rivers: Connectivity and the hyporheic corridor, *Journal of the North American Benthological Society* 12:48–60.

29. See Stanford and Ward, Ecosystem perspective of alluvial rivers. These authors give another possible cause for the wide hyporheic zones they have found in Montana:

that expanses of glacial outwash—swaths of unconsolidated gravel and cobbles washed out in the meltwater flowing from melting ice sheets at the end of the last ice age—have created a porous bed through which water flows even though the surface of the bed is now soil-covered and vegetated.

30. Ibid.

31. Stanford and Ward, Hyporheic habitat of river ecosystems. The animals that live only in the phreatic zone are known as *stygobionts* (*stygo-* derives from the river Styx, of classical mythology, beyond which lies Hades).

32. Hynes, *Ecology of running waters.*

33. W. B. White, D. C. Culver, J. S. Herman, T. C. Kane, and J. E. Mylroie, 1995, Karst lands, *American Scientist* 83:450–59.

34. F. J. Triska, J. R. Sedell, and S. V. Gregory, 1982, Coniferous forest streams, in *Analysis of coniferous forest ecosystems in the western United States,* ed. R. L. Edmonds, 292–332 (Stroudsberg, Pennsylvania: Hutchinson Ross).

35. F. J. Swanson, S. V. Gregory, J. R. Sedell, and A. G. Campbell, 1982, Land-water interactions: The riparian zone, in *Analysis of coniferous forest ecosystems in the western United States,* ed. R. L. Edmonds, 267–91 (Stroudsberg, Pennsylvania: Hutchinson Ross).

36. R. J. Naiman, C. A. Johnston, and J. C. Kelley, 1988, Alteration of North American streams by beaver, *BioScience* 38:753–62.

37. Crickmay, *Work of the river.*

38. Leopold, *View of the river.*

39. Peace-Athabasca Delta Group, 1972, *The Peace-Athabasca Delta: A Canadian resource,* Summary Report (Ottawa: Environment Canada). Also, J. Andrews, ed., 1993, *Canada water book* (Ottawa: Environment Canada); D. M. Rosenberg, R. A. Bodaly, R. E. Hecky, and R. W. Newbury, 1987, The environmental assessment of hydroelectric impoundments and diversions in Canada, in *Canadian aquatic resources,* ed. M. C. Healey and R. R. Wallace, 71–104 (Ottawa: Department of Fisheries and Oceans).

40. Rosenberg et al., Environmental assessment, 77.

41. Peace-Athabasca Delta Implementation Committee, 1987, *Peace-Athabasca Delta water management works evaluation* (Ottawa: Environment Canada).

42. The cottonwoods concerned are chiefly *Populus deltoides,* the prairie cottonwood. Other affected species are *P. angustifolia, P. balsamifera,* and *P. fremontii* (S. B. Rood and J. H. Mahoney, 1990, Collapse of riparian poplar forests downstream from dams in western prairies: Probable causes and prospects for mitigation, *Environmental Management* 14:451–64).

43. Ibid.

44. The map is adapted from one in ibid. Rood's and Mahoney's paper gives details on dam locations, and references.

45. The reciprocal of the average return period is called the *flood frequency.* For

instance, if the average return period at a site for floods of a given size is 10 years, the flood frequency is 1/10.

46. C. M. Pearce, D. McLennan, and L. D. Cordes, 1988, The evolution and maintenance of white spruce woodlands on the Mackenzie Delta, N.W.T., Canada, *Holarctic Ecology* 11:248–58.

47. A. E. Porsild and W. J. Cody, 1980, *Vascular plants of continental Northwest Territories, Canada* (Ottawa: National Museums of Canada), 57.

48. P. Marsh and M. Hey, 1992, Spatial variations in the spring flooding of Mackenzie Delta lakes, in *National Hydrology Research Institute Symposium,* series 7, ed. R. D. Robarts and M. L. Bothwell (Saskatoon: Environment Canada).

CHAPTER SEVEN

1. S. R. Carpenter, S. G. Fisher, N. B. Grimm, and J. F. Kitchell, 1992, Global change and freshwater ecosystems, *Annual Review of Ecology and Systematics* 23: 119–39.

2. The Canadian Shield (sometimes called the Laurentian Shield) is the ancient (Precambrian) block of hard, granitic and metamorphic rocks that forms the geological heart of the North American continent.

3. E. C. Pielou, 1991, *After the ice age: The return of life to glaciated North America* (Chicago: University of Chicago Press).

4. J. Bastedo, 1994, *Shield country: Life and times of the oldest piece of the planet* (Calgary, Alberta: Arctic Institute of North America, University of Calgary).

5. E. C. Pielou, 1994, *A naturalist's guide to the Arctic.* (Chicago: University of Chicago Press).

6. Environment Canada, 1988, *The Great Lakes: An environmental atlas and resource book* (Toronto: Environment Canada, Conservation and Protection, Ontario Region), 3.

7. M. J. Burgis and P. Morris, 1987, *The natural history of lakes* (Cambridge: Cambridge University Press).

8. Pielou, *After the ice age.*

9. Burgis and Morris, *Natural history of lakes.*

10. Pielou, *Naturalist's guide to the Arctic.*

11. Judging the precipitation falling on a lake from rain gauges on the shores usually gives an overestimate; any hills in the neighborhood of a lake tend to force moist air upward to cooler elevations, causing rainfall to be greater on the surrounding land than on the lake itself.

12. L. Pringle, 1985, *Rivers and lakes* (Alexandria, Virginia: Time-Life Books).

13. Environment Canada, *Great Lakes.*

14. D. W. Schindler, K. G. Beaty, E. J. Fee, D. R. Cruikshank, E. R. DeBruyn, D. L. Findlay, G. A. Linsey, J. A. Shearer, M. P. Stainton, and M. A. Turner, 1990,

Effects of climatic warming on lakes of the central boreal forest, *Science* 250:967–70.

15. More precisely, *pure* water is most dense at 3.94°C; its density is then exactly 1 gram per milliliter.

16. This is because the change in density for a given change in temperature is greater at high temperatures than at low.

17. R. E. Thomson, 1981, *Oceanography of the British Columbia coast* (Ottawa: Department of Fisheries and Oceans).

18. Ibid.

19. Ibid.

20. These numbers apply to *deepwater waves,* i.e., to waves whose wavelength is more than twice the water's depth. They are derived from Thomson, *Oceanography of the British Columbia coast,* table 7.8.

21. The description applies to *uninodal* seiches, as in figure 7.6. Binodal and trinodal seiches sometimes occur. A bimodal seiche has a wavelength equal to the lake's length, and a trimodal seiche a wavelength equal to three-quarters of the lake's length.

22. R. Wetzel, 1975, *Limnology* (Philadelphia: W. B. Saunders).

23. Burgis and Morris, *Natural history of lakes.*

24. The density of air at sea level when the temperature is 0°C is 0.001276 grams per milliliter.

25. The conditions are that the temperatures above and below the interface be 10°C and 5°C, respectively. The difference in amplitude between surface and internal waves is very sensitive to to the temperatures at the interface. See R. A. Bryson and R. A. Ragotzkie, 1960, On internal waves in lakes, *Limnology and Oceanography* 5: 397–408.

26. Wetzel, *Limnology.*

27. Bryson and Ragotzkie, On internal waves in lakes.

28. These are merely representative figures, interpolated from data in Thomson, *Oceanography of the British Columbia coast.*

29. Bryson and Ragotzkie, On internal waves in lakes.

30. Slicks are usually, but not invariably, above the troughs of internal waves. See Thomson, *Oceanography of the British Columbia coast.*

31. Ibid.

32. H. B. N. Hynes, 1970, *The ecology of running waters* (Toronto: University of Toronto Press).

33. P. R. Ehrlich, A. H. Ehrlich, and J. P. Holdren, 1977, *Ecoscience: Population, resources, environment* (San Francisco: W. H. Freeman & Co.).

34. Note the following acronyms, which are easy to confuse: *DOC* means dissolved organic carbon; *DO* means dissolved oxygen; *BOD* means biochemical (or biological) oxygen demand.

35. The vast majority of organisms require oxygen, but not all. Those that don't are known as *anaerobic*. See section 7.7.

36. D. W. Schindler, 1974, Eutrophication and recovery in experimental lakes: Implications for management, *Science* 184:897–99.

37. F. J. Swanson, T. K. Kratz, N. Caine, and R. G. Woodmansee, 1988, Landform effects on ecosystem patterns and processes, *BioScience* 38:92–98.

38. Burgis and Morris, *Natural history of lakes.*

39. The well-lit water above the compensation level is the *euphotic zone;* the dark water below it is the *profundal zone.*

40. Ehrlich, Ehrlich, and Holdren, *Ecoscience.*

41. The inshore waters around a lake where the water is shallow enough for rooted plants to grow is called the *littoral zone;* the deeper, offshore waters are the *limnetic zone.*

42. Burgis and Morris, *Natural history of lakes.*

43. Wetzel, *Limnology.*

44. M. A. Kamrin, 1996, The mismeasure of risk, *Scientific American,* September, 178–80.

45. R. L. Smith, 1980, *Ecology and field biology* (New York: Harper & Row).

46. Ehrlich, Ehrlich, and Holdren, *Ecoscience.*

47. Burgis and Morris, *Natural history of lakes.*

48. According to the International Joint Commission, a body created by Canada and the United States to advise both governments on Great Lakes water issues, the following were (as of November 1995) the most serious pollutants in the lakes: PCBs, DDT, dieldrin, toxaphene, TCDD (a dioxin, the active ingredient in Agent Orange), TCDF (a furan), mirex, mercury, alkylated lead, benzo(a)pyrene, and hexachlorobenzene (http://www.seagrant/wisc.edu/Communications/Publications/One-pagers/Critical Pollutants).

49. Environment Canada, *Great Lakes,* 30.

50. Acidity is measured in pH units. Pure, neutral water (neither acidic nor alkaline) has a pH of 7. Values less than 7 indicate acidity. The pH of unpolluted rain is about 6—the atmospheric carbon dioxide dissolved in it makes it weakly acidic. Rain in which toxic industrial gases is dissolved—acid rain—begins to be damaging when the pH falls below 5; below 4 is serious. According to Ehrlich, Ehrlich, and Holdren, *Ecoscience,* "3 to 3.5 [is] not unusual and one rain with a measured pH of 2.1 [has been] recorded" (661).

51. The chief "polluting" metals are zinc, copper, selenium, mercury, cadmium, lead, and arsenic. See R. D. Hamilton et al., 1987, Major aquatic contaminants, their sources, distribution, and effects, in *Canadian aquatic resources,* ed. M. C. Healey and R. R. Wallace, 357–86 (Ottawa: Department of Fisheries and Oceans).

CHAPTER EIGHT

1. B. Michel and R. O. Ramseier, 1971, Classification of river and lake ice, *Canadian Geotechnical Journal* 8:36–45.

2. A typical speed would be 0.2 millimeters per second, which is about 1/750 of the speed of the current produced at the surface by a gentle breeze in summer. See R. Wetzel, 1975, *Limnology* (Philadelphia: W. B. Saunders); B. C. Kenney, 1991, Under-ice circulation and the residence time of a shallow bay, *Canadian Journal of Fisheries and Aquatic Sciences* 48:152–62.

3. Wetzel, *Limnology*.

4. M. J. Burgis and P. Morris, 1987, *The natural history of lakes* (Cambridge University Press).

5. M.-K. Woo, 1986, Permafrost hydrology in North America, *Atmosphere-Ocean* 24:201–34.

6. R. Gerard, 1989, Ice formation on northern rivers, in *Northern lakes and rivers,* ed. W. C. Mackay, 23 (Edmonton: Boreal Institute for Northern Studies).

7. J. R. Mackay, 1963, *The Mackenzie Delta area, N.W.T.* Memoir no. 8 (Ottawa: Department of Mines and Technical Surveys, Geographical Branch).

8. J. Bastedo, 1994, *Shield country: Life and times of the oldest piece of the planet* (Calgary, Alberta: Arctic Institute of North America, University of Calgary).

9. Sunlight consists of electromagnetic radiation of a wide range of wavelengths. Its total energy is partitioned among short wave radiation (ultraviolet plus X rays), with 9 percent of the energy; visible light, with 41 percent; and long-wave radiation (infrared plus heat) with 50 percent. The wavelengths of visible light range from 0.4 to 0.7 micrometers; those of short-wave and long-wave radiation lie below and above these respective limits.

10. M.-K. Woo and R. Heron, 1989, Freeze-up and break-up of ice cover on small arctic lakes, in *Northern lakes and rivers,* ed. W. C. Mackay, 56–62 (Edmonton: Boreal Institute for Northern Studies).

11. J. D. LaPerriere, 1981, Vernal overturn and stratification of a deep lake in the high subarctic under ice, *Verhandlungen, Internationale Vereinigung für theoretische und angewandte Limnologie* 21:288–92. Also, J. P. Gosink and J. D. LaPerriere, 1986, Short-wave heating of lake surface water under candled ice cover, in *Cold Regions Hydrology Symposium* (Bethesda, Maryland: American Water Resources Association), 31–38.

12. T. D. Prowse, 1989, Ice break-up on northern rivers: The Liard River as an example, in *Northern lakes and rivers,* ed. W. C. Mackay, 24–42 (Edmonton: Boreal Institute for Northern Studies).

13. The "moats" around the margins of thawing lakes, described above, are sometimes formed in the same manner, in which case they have icy "floors."

14. Prowse, Ice break-up on northern rivers.

15. J. R. Mackay, *Mackenzie Delta area.*

16. *Permanently* doesn't mean *eternally.* To avoid vagueness, permafrost, also called *perennially* frozen ground, is precisely defined as ground that has remained frozen for more than one year. Usually it has been frozen for centuries.

17. J. Whittow, 1984, *The Penguin dictionary of physical geography* (New York: Penguin Books).

18. D. M. Lawler, 1993, Needle ice processes and sediment mobilization on river banks: The River Ilston, West Glamorgan, UK, *Journal of Hydrology* 150:81–114. In the arctic tundra, freezing of the ground causes rivers to erode their banks by a different process: extreme winter cold causes the soil to crack; cracks close to, and parallel with, a riverbank force blocks of soil to slump into the river. See P. J. Williams and M. W. Smith, 1989, *The frozen earth: Fundamentals of geocryology* (New York: Cambridge University Press).

19. When a given quantity of water turns to ice, it releases energy, and for this ice to melt and become water again it must absorb an equal amount of energy. The energy exchanged (in either direction) is called the *latent heat of fusion* of the water.

20. Woo, Permafrost hydrology in North America.

21. R. O. van Everdingen, 1987, The importance of permafrost in the hydrological regime, in *Canadian aquatic resources,* ed. M. C. Healey and R. R. Wallace, 243–76 (Ottawa: Department of Fisheries and Oceans).

22. R. O. van Everdingen, 1976, Geocryological terminology, *Canadian Journal of Earth Sciences* 13:862–67. Also, van Everdingen, Importance of permafrost.

23. This defines *ice-rich* permafrost. See van Everdingen, Geocryological terminology.

24. S. B. Young, 1989, *To the Arctic: An introduction to the far northern world* (New York: John Wiley & Sons).

25. Williams and Smith, *Frozen earth.*

26. Woo, Permafrost hydrology in North America.

27. The existence of suprapermafrost talik means that there is ground above the permafrost table that doesn't freeze in winter. This is a sign of recent climatic warming that has warmed the uppermost ground layers but not, so far, penetrated deep.

28. M. Church, 1974, Hydrology and permafrost with reference to northern North America, in *Proceeding of the Workshop Seminar on Permafrost Hydrology,* Canadian National Committee, International Hydrological Decade, Ottawa, 7–20. Also, van Everdingen, Importance of permafrost.

29. Van Everdingen, Importance of permafrost.

30. Church, Hydrology and permafrost.

31. W. D. Billings and K. M. Peterson, 1980, Vegetational change and ice-wedge polygons through the thaw-lake cycle in arctic Alaska, *Arctic and Alpine Research* 12: 413–32.

32. J. R. Mackay et al., 1972, Relic Pleistocene permafrost, Western Arctic, Canada, *Science* 176:1321–23.

33. J. R. Mackay, 1994, Pingos and pingo ice of the western Arctic coast, Canada, *Terra* 106:1–11.

34. The text describes the growth of *closed-system* pingos, which are the large, spectacular, truly arctic kind. Smaller ice-cored "volcanos," known as *open-system* pingos, also occur. They are found near the foot of long slopes, in regions where the permafrost is thin and patchy. They are artesian springs where groundwater is forced up from below through a hole in the permafrost and freezes into a plug of ice as it emerges.

35. Tundra polygons do not necessarily remain basin-shaped as they age. See E. C. Pielou, 1994, *A naturalist's guide to the Arctic* (Chicago: University of Chicago Press).

36. R. P. Sharp, 1988, *Living ice: Understanding glaciers and glaciation* (Cambridge: Cambridge University Press).

CHAPTER NINE

1. M. Reisner, 1986, *Cadillac Desert: The American west and its disappearing water* (New York: Viking Press), 28.

2. P. R. Ehrlich and E. H. Ehrlich, 1991, *Healing the planet* (Reading, Massachusetts: Addison-Wesley).

3. R. M. Baxter, 1977, Environmental effects of dams and impoundments, *Annual Review of Ecology and Systematics* 8:255–83.

4. M. J. Burgis and P. Morris, 1987, *The natural history of lakes* (Cambridge: Cambridge University Press).

5. T. D. Brock et al., 1994, *Biology of microorganisms,* 7th edition (Englewood Cliffs, New Jersey: Prentice Hall).

6. C. Dumont, 1995, Mercury and health: The James Bay Cree experience, *Proceedings of the Canadian Mercury Network Workshop,* 1995 (Montreal: Cree Board of Health and Social Services).

7. R. A. Bodaly, R. E. Hecky, and R. J. P. Fudge, 1984, Increases in fish mercury levels in lakes flooded by the Churchill River diversion, northern Manitoba, *Canadian Journal of Fisheries and Aquatic Sciences* 41:682–91. Also, R. A. Bodaly et al., 1984, Collapse of the lake whitefish *(Coregonus clupeaformis)* fishery in southern Indian Lake, Manitoba, following lake impoundment and river diversion, *Canadian Journal of Fisheries and Aquatic Sciences* 41:682–91.

8. J. W. M. Rudd, R. Harris, C. A. Kelly, and R. E. Hecky, 1993, Are hydroelectric reservoirs significant sources of greenhouse gases? *Ambio* 22:246–48.

9. P. R. Ehrlich, A. H. Ehrlich, and J. P. Holdren, 1977, *Ecoscience: Population, resources, environment* (San Francisco: W. H. Freeman & Co.).

10. B. F. Chao, 1991, Man, water, and global sea level, *EOS, Transactions, American Geophysical Union,* 72:492.

11. B. F. Chao, 1995, Anthropogenic impact on global geodynamics due to reservoir water impoundment, *Geophysical Research Letters* 22:3529–32. The conclusions are summarized in Reservoirs speed up Earth's spin, 1996, *Science News,* February 17, 108.

12. For a full account of one such disaster, in the Nechako valley of British Columbia, see B. Christensen, 1995, *Too good to be true: Alcan's Kemano Completion Project* (Vancouver: Talon Books).

13. Baxter, Environmental effects of dams and impoundments.

14. Burgis and Morris, *Natural history of lakes.*

15. N. D. Gordon, T. A. McMahon, and B. L. Finlayson, 1992, *Stream hydrology: An introduction for ecologists* (New York: John Wiley & Sons).

16. Christensen, *Too good to be true.*

17. R. Cannings and S. Cannings, 1996, *British Columbia: A natural history* (Vancouver: Douglas & McIntyre).

18. H. J. A. Neu, 1970, A study on mixing and circulation in the St. Lawrence estuary up to 1964, AOL Report 1970-9 (Dartmouth, Nova Scotia: Atlantic Oceanography Laboratory, Bedford Institute of Oceanography).

19. Gordon, McMahon, and Finlayson, *Stream hydrology.*

20. R. W. Newbury, G. K. McCullough, and R. E. Hecky, 1984, The southern Indian Lake impoundment and Churchill River diversion, *Canadian Journal of Fisheries and Aquatic Sciences* 41:548–57.

21. Christensen, *Too good to be true.*

CHAPTER TEN

1. T. H. Whillans, 1987, Wetlands and aquatic resources, in *Canadian aquatic resources,* ed. M. C. Healey and R. R. Wallace, 321–56 (Ottawa: Department of Fisheries and Oceans).

2. W. A. Niering, 1966, *The life of the marsh* (New York: McGraw-Hill).

3. In the Canadian prairies the annual productivity (grams dryweight per square meter) of wetlands, agricultural lands, and grasslands are, respectively, 2,000, 650, and 500. See J. Shay, 1981, quoted in C. D. Rubec, P. Lynch-Stewart, G. M. Wickware, and I. Kessel-Taylor, 1988, Wetland utilization in Canada, in National Wetlands Working Group, *Wetlands of Canada,* Ecological Land Classification Series, no. 24 (Ottawa: Canadian Government Publishing Centre, Supply and Services), 381–412.

4. T. C. Winter, 1981, Effects of water-table configuration on seepage through lakebeds, *Limnology and Oceanography* 26:925–34.

5. E. C. Pielou, 1991, *After the ice age: The return of life to glaciated North America* (Chicago: University of Chicago Press).

6. National Wetlands Working Group, 1988, *Wetlands of Canada,* Ecological

Land Classification Series, no. 24 (Ottawa: Canadian Government Publishing Centre, Supply and Services).

7. C. Tarnocai et al., 1988, The Canadian wetland classification system, in National Wetlands Working Group, *Wetlands of Canada,* Ecological Land Classification Series, no. 24 (Ottawa: Canadian Government Publishing Centre, Supply and Services), 415–27.

8. S. C. Zoltai, 1988, Wetland environments and classification, in National Wetlands Working Group, *Wetlands of Canada,* Ecological Land Classification Series, no. 24 (Ottawa: Canadian Government Publishing Centre, Supply and Services), 1–26.

9. Niering, *Life of the marsh.*

10. A. W. H. Damman, 1986, Hydrology, development, and biogeochemistry of ombrogenous peat bogs with special reference to nutrient relocation in a western Newfoundland bog, *Canadian Journal of Botany* 64:384–94.

11. Ibid.

12. P. J. Williams and M. W. Smith, 1989, *The frozen earth: Fundamentals of geocryology* (New York: Cambridge University Press).

13. D. R. Foster, G. A. King, P. H. Glaser, and H. E. Wright, Jr., 1983, Origin of string patterns in boreal peatlands, *Nature* 306:256–58.

14. Zoltai, Wetland environments and classification.

15. The map is based on M.-K. Woo, R. D. Rowsell, and R. G. Clark, 1993, *Hydrological classification of Canadian prairie wetlands and prediction of wetland inundation in response to climatic variability,* Occasional Paper no. 79 (Ottawa: Canadian Wild Service), figure 1. The boundary of the ice sheet is from Geological Survey of Canada, 1987, *Late Wisconsinan glacier complex,* map 1584A, scale 1:7.5 million (Ottawa: Department of Energy, Mines, and Resources, Survey and Mapping Branch).

16. Niering, *Life of the marsh.*

17. Ponds smaller than 0.4 hectares usually dry up, those larger than 2 hectares usually do not; for ponds of intermediate areas, it depends on the season. See G. D. Adams, 1988, Wetlands of the prairies of Canada, in National Wetlands Working Group, *Wetlands of Canada,* Ecological Land Classification Series, no. 24 (Ottawa: Canadian Government Publishing Centre, Supply and Services), 155–98.

CHAPTER ELEVEN

1. The sizes range from less than a micrometer (0.001 millimeter) to 5 or 6 millimeters. The largest are just visible to the naked eye, but a microscope is needed to see any detail.

2. Viruses are not included as they are not, themselves, living entities. They are simply "packaged" genetic material that must invade a living cell in order to multiply.

3. T. D. Brock, M. T. Madigan, J. M. Martinko, and J. Parker, 1994, *Biology of microorganisms,* 7th edition (Englewood Cliffs, New Jersey: Prentice Hall). For the

history of the current classification, see V. Morell, 1997, Microbiology's scarred revolutionary, *Science* 276:699–702.

4. The traces are *stromatolites,* globular, layered rocks that look like "stone cabbages." The layers consist of sediments trapped between layers of prokaryote cells. Stromatolites are still being formed by living cells today, showing how the fossil ones must have developed. See S. J. Gould, 1989, *Wonderful life: The Burgess shale and the nature of history* (New York: W. W. Norton & Co.).

5. R. Buchsbaum and L. J. Milne, 1967, *The lower animals* (Garden City, New York: Doubleday & Co.).

6. G. E. Hutchinson, 1967, *A treatise on limnology,* vol. 2, *Introduction to lake biology and the limnoplankton* (New York: John Wiley & Sons).

7. The Cyanobacteria used to be called *blue-green algae* and are so called in all but recent books. Two now outdated technical names for them are *Cyanophyta* and *Myxophyceae.* The blue pigment in the majority of them is *phycocyanin.* See Brock et al., *Biology of microorganisms.*

8. W. T. Edmondson, ed., 1959, *Freshwater biology,* 2d edition (New York: John Wiley & Sons).

9. D. W. Schindler, 1974, Eutrophication and recovery in experimental lakes: Implications for management, *Science* 184:897–99.

10. Green protists are also known as unicellular algae, a name that conceals their separateness from the multicellular algae, which are the simplest and most primitive plants.

11. Hutchinson, *Treatise on limnology.*

12. Edmondson, *Freshwater biology.* Diatoms are described in a chapter titled "Bacillariophyceae," the scientific term for diatoms.

13. Hutchinson, *Treatise on limnology.*

14. Some classificatory details: the water fleas belong to the order Cladocera, and the copepods to the order Copepoda, in the class Crustacea (which also contains crabs, lobsters, shrimps, and their ilk) in the phylum Arthropoda.

15. Buchsbaum and Milne, *Lower animals.*

16. M. J. Burgis and P. Morris, 1987, *The natural history of lakes* (Cambridge University Press).

17. Ibid.

18. Edmondson, *Freshwater biology.*

19. Animals that can swim where they wish, independent of water turbulence, are collectively known as *nekton.* Swimmers like water fleas and copepods, too small to resist turbulence, are on the borderline between plankton and nekton and are sometimes called *nektoplankton,* though most scientists treat them as plankton. See Hutchinson, *Treatise on limnology.*

20. Ibid.

21. D. W. Schindler, K. G. Beaty, E. J. Fee, D. R. Cruikshank, E. R. DeBruyn,

D. L. Findlay, G. A. Linsey, J. A. Shearer, M. P. Stainton, and M. A. Turner, 1990, Effects of climatic warming on lakes of the central boreal forest, *Science* 250:967–70.

22. The inorganic compounds are ammonium salts and nitrates. A small amount of nitrogen fixation happens in lightning strikes and in industrial furnaces.

23. P. R. Ehrlich, A. H. Ehrlich, and J. P. Holdren, 1977, *Ecoscience: Population, resources, environment* (San Francisco: W. H. Freeman & Co.).

24. Schindler, Eutrophication and recovery.

25. Brock et al., *Biology of microorganisms.* See also Edmondson, *Freshwater biology.*

26. Brock et al., *Biology of microorganisms.*

27. J. L. Bubier, T. R. Moore, and S. Juggins, 1995, Predicting methane emissions from bryophyte distributions in northern Canadian peatlands, *Ecology* 76:677–93.

28. Burgis and Morris, *Natural history of lakes.*

29. J. W. M. Rudd et al., 1993, Are hydroelectric reservoirs significant sources of greenhouse gases? *Ambio* 22:246–48.

30. S. R. Carpenter et al., 1992, Global change and freshwater ecosystems, *Annual Review of Ecology and Systematics* 23:119–39.

31. Rudd et al., Hydroelectric reservoirs.

CHAPTER TWELVE

1. Any gas can be made to condense into a liquid if the pressure upon it is increased sufficiently and the temperature reduced below a critical level. A vapor is a gas that, at ambient temperatures, will condense under increased pressure alone.

2. C. W. Fetter, 1988, *Applied hydrogeology* (New York: Macmillan).

3. Ibid. Also, R. Brassington, 1988, *Field hydrology* (New York: Halsted Press).

4. R. C. Ward, 1967, *Principles of hydrology* (New York: McGraw-Hill).

5. Fetter, *Applied hydrogeology.*

6. C. W. Thornthwaite and J. R. Mather, 1957, *Instructions and tables for computing potential evapotranspiration and the water balance,* Publication 10 (Centerton, New Jersey: Laboratory of Climatology), 185–311, cited in Fetter, *Applied hydrogeology,* 33.

7. Condensation of water vapor into ice is sometimes called *sublimation,* the term also used for the conversion of ice to vapor. It is less confusing to speak of the vapor-to-ice conversion as *condensation*—the word that also means the change from vapor to liquid.

8. V. J. Schaefer and J. A. Day, 1981, *A field guide to the atmosphere,* Peterson Field Guide 26 (Boston: Houghton Mifflin Co.).

9. D. Brunt, 1942, *Weather study* (New York: Thomas Nelson & Sons).

10. Schaefer and Day, *Field guide to the atmosphere.*

11. Equivalently, at low temperatures, the saturation vapor pressure of water vapor is less with respect to ice than it is with respect to liquid water.

12. Brunt, *Weather study.*

13. G. Petts and I. Foster, 1985, *Rivers and landscape* (London: Edward Arnold).

14. H. Miller, 1977, *Water at the surface of the earth* (New York: Academic Press).

15. B. F. McKay and B. F. Findlay, 1971, Variation of snow resources with climate and vegetation in Canada, in *Proceedings of the 39th Western Snow Conference,* 17–26, cited in Miller, *Water at the surface of the earth.*

16. Snow gauging is commonly done with a 4-centimeter-diameter aluminum tube, separated into meter-long sections; these are added one atop another as each preceding one sinks below the snow surface; the starter tube has a sharp steel edge.

17. I. C. Strangeways, 1985, Automatic weather stations, in *Facets of hydrology II,* ed. J. C. Rodda, 25–68 (New York: John Wiley & Sons).

18. R. D. Harr, 1981, Some characteristics and consequences of snowmelt during rainfall in western Oregon, *Journal of Hydrology* 53:277–304. See also R. D. Harr, 1986, Effects of clearcutting on rain-on-snow runoff in western Oregon: A new look at old studies, *Water Resources Bulletin* 22:1095–1100.

Index

Page references for definitions appear in italics.